TINY HOUSE DESIGN & CONSTRUCTION GUIDE

YOUR GUIDE TO BUILDING A MORTGAGE FREE, ENVIRONMENTALLY SUSTAINABLE HOME

2nd Edition

Dan Louche

Photography: Dan Louche
Graphic Design: Aleksandar Doskovic

CONTENTS

It is extremely important that you use the safest methods possible while building your tiny house

BE SAFE!

—

Below are a few reminders while working on any project:

- *Always use caution, good judgment, and common sense when following the procedures described in this guide or elsewhere.*

- *Read and follow any instructions or warning labels on both products and tools as they take precedence over any instructions in this guide.*

- *Special care should be taken when working with power tools. Power tools are much faster than your reaction speed, so you will need to put distance between the moving parts of the tools and yourself. If you need to hold a piece of wood closer to a blade, for instance when working with smaller pieces, consider cutting a new piece or using a clamp to hold the wood instead of your hand. Keep your hands as far away from any blades as possible and do not wear loose fitting clothing.*

- *Always wear eye and hearing protection, especially when working with power tools or using a striking tool like a framing hammer or sledge hammer. Additional personal protective equipment (PPE) is also recommended, depending on the activities you are undertaking.*

The most important piece of advice I can give when it comes to safety is to take your time and do not rush. Almost every injury I have witnessed on a jobsite is due to either carelessness, distraction, or trying to work too fast. Give yourself extra time to complete every task. If you realize that you are not going to get something done as quickly as you anticipated, which will happen, just accept it and readjust your completion schedule to a later time. Do not try to speed up.

Also, try to avoid working with large quantities of time sensitive materials at one time. For example, construction glue begins to harden 10-20 minutes after it has been applied. If you dispense too much of it and then are not able to complete the task for which it has been used, you may find yourself rushing. This is the exact scenario that caused me to injure myself early on in my construction career. In an attempt to be more efficient, I applied construction adhesive to a section of wall framing larger than I could complete in 10 minutes. Instead of just stopping and removing the glue, I sped up and became careless, resulting in a minor injury to my hand.

Take your time, be careful, and think through what you are going to do before doing it.

Building your own tiny house is an
achievable dream!

INTRODUCTION

—

0

My Mom's House

In August 2009 I received an unsettling call from my mother. The poorly constructed trailer home she was living in was beginning to deteriorate around her. Water lines had been leaking for some time and mold was growing rampant. Living under these conditions was causing her health to deteriorate, but neither she nor I had the money to purchase a conventional house or even a new trailer home. So I started researching our options. I had always been interested in smaller homes, but up until this point, I had no idea there was an entire movement around tiny living. Once I discovered it, I was hooked. I began building my mother a tiny house of her own in September 2009.

After the house was complete and my mother had moved in, I was amazed by the level of joy that it brought her. Her excitement was contagious as others who had previously been skeptical of tiny living were now genuinely considering the possibility of living in a tiny house themselves. When I saw this reaction, I knew that I wanted to help others experience a similar level of happiness and independence. I founded Tiny Home Builders and have been sharing my knowledge on tiny houses ever since.

I imagine, since you are reading this book, you too are excited about the possibilities that a tiny house can bring; the financial freedom of not having a mortgage, the freedom to move as you desire and to take your house with you, and finally the freedom of a simpler life. I hope you find answers and inspiration in these pages and realize that building your own tiny house is an achievable dream.

Before beginning construction, you
will want to start preparing to make
the transition to your new house

PREPARING
TO LIVE TINY

—

DOWNSIZING

To be comfortable in a tiny house you are likely going to have to downsize your belongings. It is better to do this well before moving into your new house so that you have time to purposefully decide what you will keep and what you will get rid of. Downsizing is not something that should be done as part of your move. It is a journey and not a weekend task. This process should be started at least 3 months prior to moving, but there is no need to wait until then. There are many benefits to downsizing other than just being able to fit into your new home.

This process can be easier for some and harder for others, but since you are considering a tiny house you are likely already inclined to living with and wanting less.

The first step to downsizing is to identify what is important to you and what makes you happy. If there is something that you use everyday or if it brings you joy, then you will need to find a place for it in your new home. Do not abandon the activities that make you happy because you do not think they will fit. For other items that are used less frequently, you will need to evaluate how much value they provide to your life. This process can be difficult since you will need to differentiate between actual value and perceived value.

Hobbies and all of the supplies that come with them are a great example of where some of these difficult decisions may arise. If you have a hobby that you actively participate in and really enjoy, then you will want to carve out space for it in your new home. However, just having the supplies for an activity does not make it your hobby. Often times, people will hang on to belongings that represent who they want to be or what they would do if they had more time. They hold on to these things so that they are ready to become who they want to be, but often that never comes and the stuff becomes a source of stress rather than inspiration. These items act as a reminder of all the things they never had time to do or learn.

The easiest things to get rid of when downsizing is the stuff you do not use. We all have items that we either used in the past or purchased because we thought we would use them, but never did. If you do not use it, you do not need it. You may try to convince yourself that you will indeed use it someday and that is why it is worth hanging onto.

So, how do you identify what you no longer use? For most stuff, if you are honest with yourself, it should be pretty straightforward. There are, however, some items that are just harder to track. With clothing for instance it can be hard to remember the last time you wore something. A simple technique for this is to hang all your clothes backwards in your closet so that the open part of the hook on the clothes hanger is facing you. Then every time an item is worn and washed you hang it back up with the hook turned in the conventional direction. After a certain period of time, you identify what has not been worn by looking at the hook direction. Since what we wear is seasonal, this process can take up to a year.

Clothes Hung Backwards

For items other than clothes the idea is the same, but the implementation is different. You want to somehow mark the items and then unmark them as they are used. For instance, you can place small yard sale stickers inconspicuously on items around your house. When an item is used, remove the sticker. After a certain period of time, say a month or two, anything that still has its sticker is a candidate to get rid of.

Also consider the value and how easy an item is to replace. If an item is inexpensive and easily found, less thought and consideration can be given when deciding if it should be kept.

Items with sentimental value are the hardest to let go. While they are rarely used, they represent a memory that you do not want to give up. The important thing to remember here is that you are really trying to hold on to the memory and not the item itself. So, if you have inherited a collection of something, you probably do not need to own the entire collection to rekindle that memory. Instead you could keep just a single item and actually display it, instead of having it packed away. Another option is to take a picture of the items. Just seeing them in a photo will have similar effects to looking at the object in person. Simply owning something and having it packed away is not honoring the memory.

People have a tendency to expand to fill the space they are given. That is why many people, regardless of the size of the space they are in, always seem to want a bigger place. This is because most of us are consumers, and we generally add to our possessions faster than we are getting rid of our stuff. This is especially easy to do when you have excess room. If you move into a bigger place that has an extra room, you go out and buy furniture to fill that room. If your kitchen is bigger and you have more cabinets, you do not think twice about buying a single use small appliance because you can just stick it in an empty cabinet. This behavior, if unchecked, continues until you begin to feel cramped in your living space and, before long, you are dreaming about your next roomier house.

A way to stop and reverse this behavior is to artificially limit your space. It is not realistic to move to a smaller house every few months while you downsize your belongings. You can, however, close off rooms in your house or designate them as donation or sell rooms, where after a specific period of time everything in the room is taken for donation or sold. By doing this, you are limiting your space. This serves two purposes: you have less to expand into and you make a temporary holding area for stuff you plan to get rid of. By slightly delaying getting rid of your stuff, you can reduce the anxiety that you may feel about needing something the moment you get rid of it.

Make sure to follow through with actually clearing out the room and donating or selling the items. It is best if you can have a friend, family member, or charity take the stuff for you. That way, you can schedule the pickup in advance, ensure it happens, and avoid seeing or touching the items which could trigger a change of heart. If you do not have an entire room to close off, you can do the same thing with a closet or even a kitchen cabinet. You might consider designating a donation or sell cabinet just for kitchen stuff, even if you have a room or closet. It is convenient to have that cabinet nearby considering how much the kitchen is used.

A more aggressive version of this technique involves putting all of your items in the designated area and only taking things out as they are needed. This would ensure that you are not accidentally omitting something by never adding it to the designated area in the first place.

FINDING A PLACE TO PARK AND LIVE

There are a lot of options when it comes to parking your tiny house. However, living in it is a different story. Most jurisdictions consider tiny houses to be RV's, and RV's generally cannot be lived in full-time. As tiny houses gain in popularity, some counties are changing the zoning laws to be more tiny house friendly. The most common places people park their tiny houses are as follows:

One option where people have had a lot of success parking their tiny houses are RV parks. These locations usually have all the hookups needed for a tiny house and the plots are sized perfectly. Some RV park owners are recognizing the potential of tiny houses and are designating special areas just for the homes to create a more community feel.

A downside to RV parks, assuming they have not established a separate tiny house community, is that they normally have a more transient atmosphere, with limited full-time residents. They typically cater to vacationers and are not located near job centers. Finally, some RV parks will require you to move to a different plot every few months to comply with local laws.

RURAL LAND

Another option is to buy or rent rural land or non-rural land that is more wooded and less visible. While living in a tiny house may still not be legal on the land, it is very unlikely that anyone would notice or care. The biggest challenge with this option is getting utilities on the property if they are not already available.

Tiny House in an RV Park

A good way to be closer to town and your work is to live in someone's side yard or backyard. This can be a great arrangement for a host who is looking for additional income, without the loss of privacy. The most important thing here is for you and your host to communicate with your future neighbors so they understand what you are doing. A simple introduction and explanation can prevent a neighbor from calling code enforcement on you due to a misunderstanding.

You will want to find a location for your tiny house before you start building, if possible. If you plan to take a longer period of time to build your house, this can be challenging because of availability changes, but it is worth the effort to secure a location. You do not want to spend time and money constructing your dream home, only to find out that you have no place to live in it.

FINANCING AND INSURANCE

It may be challenging to find companies that will either finance or insure a tiny house. These companies' business models involve having a very good understanding of the value of the items they are financing and insuring. Tiny houses are a relatively new concept. Unlike the traditional housing industry, where prices, long term values, and rates are well defined, much of that can vary drastically or is unknown with tiny houses. This is especially true for houses built by individuals. A house built incorrectly, with fundamental mistakes, may be considered worthless.

As the popularity of tiny houses has grown, some smaller companies and even larger companies are starting to get in on the action. Financial companies like USAA and Suntrust, in the USA, are now offering loans for tiny houses and insurance companies are offering specialized policies. Be sure to check out the resources section at the back of this book for additional details.

Before any construction can begin
on your tiny house, you will need to
decide on a design

DESIGN & PLANS
—

The design that you select for your house can either be your own custom design, an existing design, or a combination of the two.

CUSTOM DESIGN

Coming up with your own custom design allows you to create a house to your exact specification. The size restrictions imposed on tiny houses on wheels can be restrictive, but can also make them easier to design. If you were given a blank page and asked to design your perfect house, you may feel pretty intimidated. Instead, if you were given a specifically sized box and told to arrange and fit specific items within it, this would feel a lot less daunting. Sometimes having too many choices can be crippling.

The difficult part of designing your own house from scratch is that it requires a lot of knowledge that you may not already possess and that may take a lot of time to acquire. You will need to know how to best take advantage of small spaces, proper framing and building techniques, and how to use the technology that can capture and document your design.

DESIGNING FOR SMALL SPACES

The key to a successful interior design in a tiny house is having the space and storage for all your belongings, while also having an open, roomy feel. Achieving these two seemingly conflicting goals in such a small space can be overwhelming, especially if you are currently living in a conventional home that has all the belongings that typically go with it. If you have already began downsizing to remove the excess, you will find this process easier.

HOUSE LAYOUT

How you lay out your house is going to determine how open and how functional it is. You will want to research and experiment as much as possible before finalizing the design.

A great resource for tiny house layouts is the RV industry. While tiny houses are relatively new, RV's have been around for over 100 years. In that time, the various manufacturers have experimented and perfected different layouts based on customer feedback. While the goals of someone who wants to go camping and the goals of someone who wants to live in a structure full-time may not perfectly align, there is still a lot that can be learned. It would be well worth the time to look through the various RV floorplans and even go to an RV dealer or RV show and see them firsthand. In doing so, you will see commonalities in many of the designs that should be considered for your house. For instance, you may notice that all of the doors to the living areas on RV's are located on the passenger side of the vehicle. This is so that when they are parked on the street, people enter and exit at the curb and not into traffic. Due to this consistency in design, all RV parks are expecting the doors to be on the passenger side and the utility hookups to be on the opposite side. If you buck this trend and end up parking your tiny house in an RV park, you may find that your front door is located right in front of the sewer hookup at the park.

As tiny houses are becoming more popular, these new designs can be a great resource. Do not just look at the pictures though. Delve deeper into the interviews and blogs, where tiny house owners discuss what they do and do not like about their house design. There will surely be things that sounded like a great idea during the design stage, but turned out to be not as practical or functional. This is especially true for the participants of tiny house TV shows. Many of their design decisions are not made by the homeowners and are purely for visual impact. The downsides to these elements would never be brought up on the show, but I have heard more than one participant say they would have done several things differently.

One of the most significant decisions that is made when designing a tiny house is the location of the bathroom and the kitchen. While some choices concerning a house's layout can be changed after the house is built, like a cabinet's location or the need for an additional bookcase, the location of the bathroom and kitchen cannot. This is because the windows and plumbing for these rooms are uniquely positioned making them very difficult to relocate. Below are two common layout choices and some pros and cons of each.

BATHROOM ALONG SHORT WALL

The bathroom fits very well along the short wall of a tiny house. If your shower and toilet are across from each other and they measure 36 inches and 28 inches respectively, that leaves a good amount of open space to move around comfortably between the two.

If the kitchen is located directly outside the bathroom, the countertop can be split between the two sides of the house creating a galley kitchen. This is an efficient kitchen design as it requires less movement while preparing meals. It also opens the kitchen up to the rest of the house which can make the whole house feel larger.

The downside is that some people prefer a longer countertop in their kitchen rather than two shorter countertops, regardless of how open it is to the rest of the house.

BATHROOM ALONG LONG WALL

Having the bathroom along the longer wall of the house and the kitchen across from it is another option. If you are someone who prefers a longer continuous countertop in your kitchen, then this may be a better option for you.

One downside to this layout is that it may feel more confined. Since the kitchen will be between the exterior wall and the bathroom, it will not feel as open. Assuming that your shower is 36 inches wide, the interior wall thickness is 4 inches, and the countertop is 25

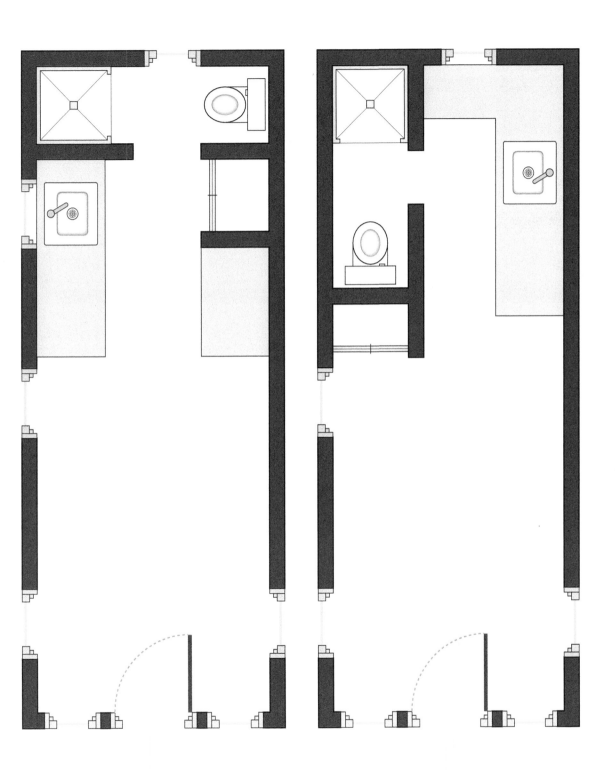

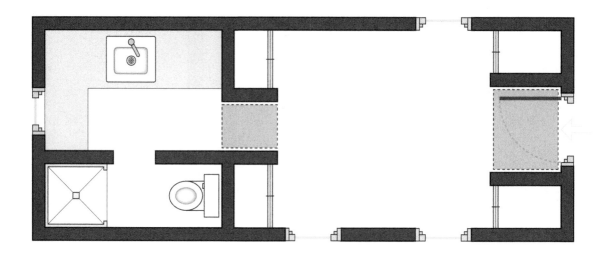

Micro-Hallways in House Design

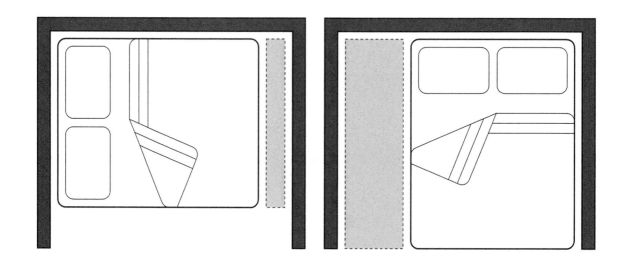

Efficient and Inefficient Furniture Placement

inches, then a tiny house with 7 feet of interior width would only have about 19 inches between the interior wall and the countertop. That might feel tight while working in the kitchen. To gain some additional room you could get a smaller shower and build a more narrow bathroom. Alternatively, you could make your countertops narrower and install a smaller sink. A slimmer countertop would also require a shallow refrigerator if placing it under the countertop. Either way, compromises may need to be made with this configuration.

When deciding on the bathroom location, you should also consider the location of the fenders on the trailer. On wider houses, it is common for the fenders to enter the living area slightly. If your bathroom overlaps the fenders, then your shower may need to be built around them. This would limit the use of some shower materials such as a one piece fiberglass shower.

MICRO-HALLWAYS

You will want to avoid including a micro-hallway in your tiny house design. It is very easy to identify a hallway in a traditional house. It is an interior passage whose purpose is to provide a way to get from one room of a house to another. A micro-hallway is a little more difficult to spot. It is any area of a house that would likely only be used to get from one part of a tiny house to another. It is essentially a hallway without the enclosure. You might be asking yourself, "who would put a hallway in a tiny house?", but they are more common than you might think. In some cases they are built into the design and structure of a house, and in other cases they are created by the placement of furniture. When designing your tiny house, try to minimize any floor space that would only be used to pass through or stood in for a few moments.

The areas indicated on the diagrams to the left are examples of micro-hallways.

FRAMING AND BUILDING BASICS

Once you have your layout and floorplan, you will need to create plans that you can build from. To do this, you will need to understand the proper way to frame a house so that your plans are structurally sound. While this can be extremely complicated in larger structures, it is much less difficult for a tiny house.

STRUCTURAL LOADS

There are forces or 'loads' that a tiny house encounters and must withstand. Some of these loads are the result of gravity, while others are caused by wind or movement, such as when the house is towed. If your house cannot handle these loads, it can experience a failure that results in a collapse.

In a 'stick built' tiny house, the framing is made up primarily of 2x4 (said 'two-by four') lumber. It is important to note that while they are called 2x4's, the dimensions are actually

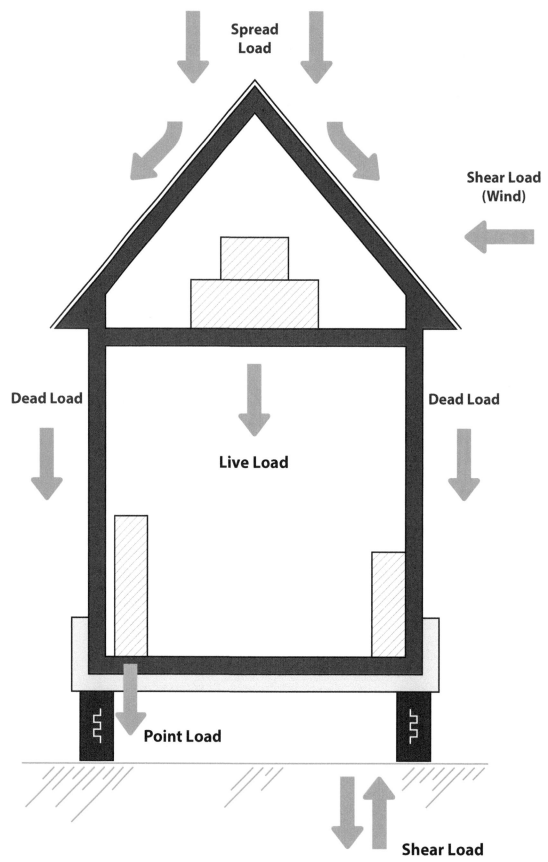

Loads on a House

1 ½ inches by 3 ½ inches (see the chapter on Materials for additional information). When a 2x4 is stood up in a vertical orientation it is very strong and efficient at supporting loads from above. When used in a wall, these 2x4's are referred to as 'studs'. For a stud to be efficient at properly carrying a load, it needs to be consistent from the top of the wall to the bottom. This is called the load path.

If a stud cannot be consistent from the top of the wall to the bottom, for instance when a window or door requires a break in the stud, then the load path needs to be redirected around the opening. This is done through the use of a header. A header is a structural support made up of 'two-by' lumber that bridges an opening. The size of the span that needs to be bridged determines the size of the lumber that the header is constructed of. See the table below for the proper sizing.

SPAN TABLE FOR HEADERS

Lumber Size	Maximum Span
2x4 (2)	48 in.
2x6 (2)	72 in.
2x8 (2)	96 in.
2x10 (2)	120 in.
2x12 (2)	144 in.

Supporting roof only, No. 2 grade lumber

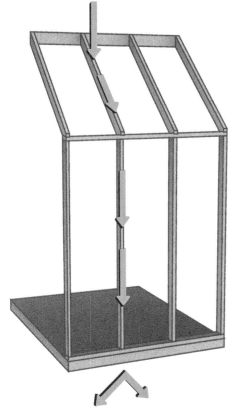
Load Path

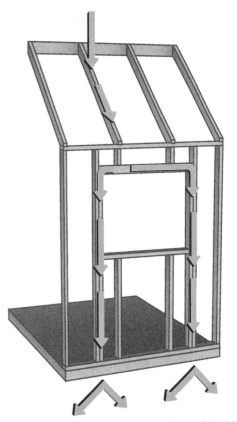
Rerouted Load Path

Headers above openings are only required in load-bearing walls. We will cover how to construct a header in the chapter on Framing.

If the load path is not consistent, then deflection can occur. Deflection is when the framing members bend under a load. If the deflection is too much and the framing member fails, then the structure can partially or fully collapse. Even a small amount of deflection can cause problems like cracks in walls, bulging windows, and difficult to open doors and windows.

The diagram to the right illustrates how studs are used in a wall to provide support.

LOAD-BEARING CANTILEVERS

A tiny house is often wider than the trailer that it is built on. When this occurs, the house subfloor will cantilever, or overhang the edge of the trailer. Because the subfloor then supports the walls of your house, it is considered a load-bearing cantilever. Load-bearing cantilevers should extend or overhang no more than the depth of the floor joist.

ADVANCED FRAMING TECHNIQUES

In recent years, framing methods have improved to make houses more energy efficient. These changes are called 'Advanced Framing Techniques'. The use of the term 'Advanced'

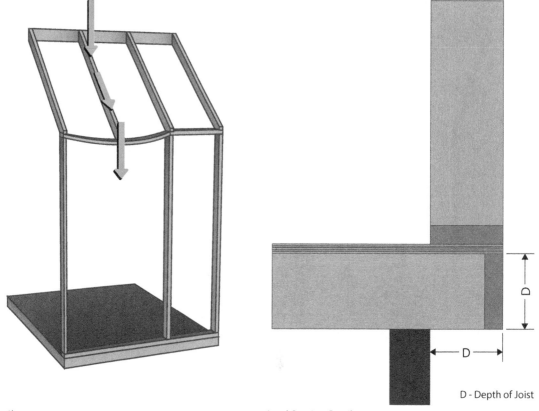

D - Depth of Joist

Deflection   Load-Bearing Cantilevers

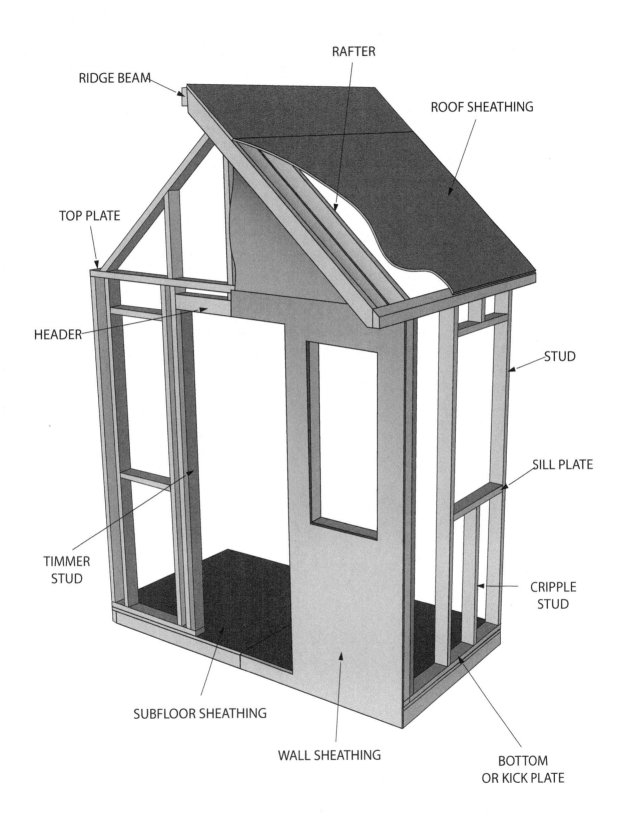

RIDGE BEAM

RAFTER

ROOF SHEATHING

TOP PLATE

HEADER

STUD

SILL PLATE

TIMMER STUD

CRIPPLE STUD

SUBFLOOR SHEATHING

WALL SHEATHING

BOTTOM OR KICK PLATE

is a bit misleading, as these methods are no more difficult to implement than 'Standard' framing. They are new ways to frame that reduce the amount of wood in a wall, increase the amount of insulation, and ensure that no 'pockets' are created that can be difficult to insulate. While not all Advanced Framing Techniques are applicable to tiny houses, there are some significant improvements that do apply, including:

24 INCH ON-CENTER STUDS. Studs in a house are generally placed either 16 or 24 inches apart from their centers (called on-center). Since the wood that studs are made of is not a good insulator, it is desirable to have the minimum number of them in your wall as possible while still providing sufficient support. For a smaller structure like a tiny house, 24 inches apart is adequate.

SINGLE TOP PLATE. Older framing methods utilize two top plates in a wall to support the weight of any rafters that may be located between the studs. The two top plates act as a form of header. However, by lining up the rafters so they make contact with the top plate at the same location the studs make contact, the weight of the roof is transferred directly to the studs and only a single top plate is required.

IMPROVED CORNER DESIGN. In older framing, the corners involved more wood and the way they were constructed actually created a hidden pocket that, unless drilled into and injected with insulation, would often go uninsulated. The new design uses less wood and is easily insulated.

3 Stud Corner   4 Stud Corner

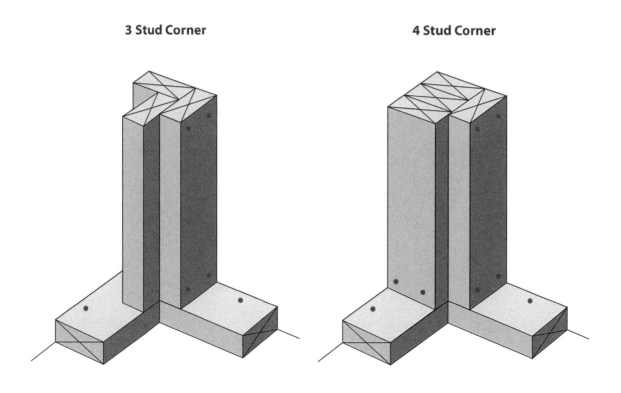

Framing Corners

OPTIMIZED WINDOW PLACEMENT. If possible, position windows so that one side of them is against an existing stud. Then, only one additional stud needs to be framed around each window. This can be difficult to do on a tiny house, since there is not normally a lot of flexibility in a window's placement.

REDUCED WOOD AROUND OPENINGS. In older framing, more wood was used as support around window and door openings than was required. Load tests show that much less wood than was traditionally being used is necessary to support the loads in these areas.

For additional information on proper building and framing standards, consult the International Building Code (IBC). For additional information on the most energy efficient framing techniques, consult the Advanced Wall Framing technology fact sheet published by the United States Department of Energy.

ALTERNATIVE FRAMING MATERIALS

There are alternatives to conventional 'stick built' or wood stud framing, including steel studs and structurally insulated panels (SIP).

STEEL STUDS

The advantages of using steel studs over wood studs makes them a compelling option for framing a conventional home. For tiny houses, however, the negatives often outweigh the benefits.

The biggest downfall of metal framing is that screws are required to attach the interior siding. For a conventional home, where drywall is used, this is not an issue. However, in tiny houses drywall is not typically used and most alternatives require nail fasteners. To get around this, people have sheathed both the exterior and the interior of the walls with plywood to provide a nailing surface for the siding. This workaround is not ideal because it is heavy, expensive, and labor intensive.

Even when a house is framed with steel studs, it is common for the kitchen and bathroom walls to still be framed with wood in order to provide a stronger mounting point for the cabinets. Considering the relative size of the kitchen and storage cabinets in a tiny house, wood framing or SIPS are recommended over steel framing.

STRUCTURALLY INSULATED PANELS (SIPS)

SIPs consist of a solid foam core insulation, sandwiched between two sheets of structural facing, typically oriented strand board (OSB). The panel core is solid and does not contain framing members, so there is no wood to act as a thermal bridge. This makes SIPs one of the most thermally efficient wall systems available.

SIPs are made by a manufacturer off-site, based on your plans. Even if your plans are not specifically made for SIPs, the manufacturer can still use them to design your walls. The downside to this approach is that any problems or issues with the panels can cause a significantly delay while replacement panels are manufactured and delivered.

Another consideration when using SIPs are the electrical and plumbing systems. The panel manufacturer can add 'chases' or channels for the wires so the switch and outlet locations will need to be provided to them as well. The plumbing should not be run through the interior of the SIPs. Plumbing will either need to be run through the floor or strips of wood will need to be placed along the interior of the wall to create a surface chase.

The cost of SIPs is normally more than conventional wood framing, but the labor savings and long term efficiency make this a worthwhile option.

CAPTURING AND DOCUMENTING YOUR DESIGN

There are several different options to capture and document your design. These range in price from free to thousands of dollars. Since the free tools that are available are more than adequate for this job and are easier on your budget, I will focus on them.

The most obvious of these options is a pencil and paper. When starting from scratch, I highly recommend beginning with a simple sketch. A sketch is great for quickly capturing

SIP Sample

your ideas and is very easy to make changes to, unlike a more complicated model in a design program. The sketch can simply be a floor plan that can be used to determine where the door(s), windows, bathroom, and kitchen should be. Try to keep the items in your sketch as close to scale as possible. I have seen people who have fit a tremendous amount of furniture and cabinets into their design by drawing things smaller. While that may look great on paper, it does not translate to real life unless you plan to use miniature furniture in your finished house.

The sketch can also be used to help determine the external appearance, since the interior design will dictate the window placement.

While the entire design can be captured and built from hand drawings, there are significant advantages to using a design and modeling computer program to convert your sketches into plans. The program that we recommend and use ourselves is Trimble SketchUp. This program has several different versions, and at the time of publishing this book, the basic personal use version is free.

Trimble SketchUp is a 3D modeling program that is incredibly easy to learn; however, if you open up the program and start with a blank canvas it can be intimidating. I recommend that you start with a sample model of a tiny house to get a feel for how the model should be constructed. These can be found online for free and are included with several paid plan packages, including the ones offered by Tiny Home Builders. By using a program that models your house in 3D, you get a unique perspective and get to see how all the different components fit together at scale. It is highly beneficial to see your design in 3D from an aesthetics perspective too. There have been several times where I sketch something and think it will look great, only to model it in 3D at perfect scale and discover that it does not look as good as I initially thought.

More importantly, having a 3D model of your house allows you to instantly determine measurements of any components in the house during construction. This can be very helpful, since architectural drawings and plans do not show every measurement. Many measurements must be derived by adding or subtracting others.

EXISTING DESIGN

Because of the size constraints, the options for the layout of a tiny house are limited. For instance, the front door can only be placed on either the back of the trailer or on one of the sides. If it is placed on the side of the house it will most likely be off-center because of the trailers fenders. If it is placed on the back of the house it will either be centered or along one of the edges. With these limitations, there are fewer design combinations than with traditional houses. The benefit is that there is a good chance a set of plans already exist for the general design you have in mind.

Settling on an existing design is the easiest, least time consuming, and most affordable option. With existing plans, there is no need to learn how to use a new design program or to be concerned if the house is framed properly and structurally sound. All of that has already been done for you. Depending on how you value your time and the cost of the plans, this option can be significantly cheaper than designing a house yourself.

Another advantage to using an existing design, particularly if at least one house has been built from the plans, is that many of the kinks have likely already been worked out. Often a design may look good on paper, but once it is built, it may not be as practical or as functional as initially believed. For instance, the first house that was built by Tiny Home Builders included a storage loft above the bathroom. Since it was small and would not carry a very large load, 2x4s were used instead of 2x6s for the loft joists. This ended up being a poor decision as few recessed lights or bathroom exhaust fans are designed to fit in a 3½ inch ceiling cavity. In this particular case, the fix was not very difficult, but it still took time and would have gone undiscovered in the plans had the house not actually been built.

Another benefit to buying plans for a house that has already been built is that you get to see what it will look like when it is completed. Just as designs may not be as functional on paper, they may not look as good in real life either.

Finally, another benefit is that some plans come with additional valuable information like a materials list. In the Building Materials chapter, we will discuss how this can save a significant amount of time and money, essentially reducing or negating the cost of the plans.

The disadvantage of using an existing design is that it may not be exactly what you want, and depending on how easy it is to modify, compromises may need to be made.

CUSTOMIZING AN EXISTING DESIGN

If you are only able to find a design that is close to what you want, but not exact, you may be able to customize it to fit your exact needs. Most plans are primarily framing plans, so changes to the exterior, including window and door locations, will require the most rework. Interior changes that work with the existing window and door placement, may not require any changes to the plans at all.

EXTERIOR CHANGES

Some minor framing and exterior customizations can be made to a design without making changes to the plans. These changes can be done during a house's construction and are called 'site modifications'. An example of this type of change might be the removal of a window or even a slightly more difficult alteration, like shifting the location of the cutout in the subfloor for the fenders. While not having the plans exactly the way you want them before you begin construction is not optimal, with care and special attention they can still work.

If the plans that you purchase are provided to you in a format that can be easily modified, you will have a lot more flexibility with the changes that you can make. For instance some plans are only available in print or PDF format, which is not modifiable. To make changes to these plans, the design would need to be recreated in a drafting program and then changed. If the plans you purchase include an AutoCAD or SketchUp file, changes can be applied with minimal effort and time. If your intention is to modify a design, inquire about receiving the plans in a change friendly format.

INTERIOR CHANGES

The framing of a house is primarily linked to the layout of the house through the window and door placement. For instance, if a kitchen is anticipated to be in a certain location, it will likely have a window positioned above the sink. If you decide to move the kitchen to another wall, you will need to ensure that any windows on that wall will not interfere with the cabinets. Most interior changes, other than those to the kitchen or bathroom, are simple to make and will not require formal changes to the plans.

Having the right tools for a job is extremely important, as they can save a considerable amount of time and frustration

TOOLS
—

0

Tools serve two main purposes. First, they make you much more efficient at performing a specific task. For instance, a pneumatic nail gun allows someone to assemble a wall much faster than someone using a hammer, while a hammer allows someone to build a wall faster than someone without one. Second, tools increase the quality of the work performed. For example, there are tools designed specifically to hide the fasteners used to assemble cabinets so they look much more professional.

There are countless tools available that can help you build a house. The problem for builders of tiny houses is that they are usually trying to minimize their belongings, so the prospect of acquiring a bunch of tools is unappealing. Not to mention the expense of obtaining all those tools. The good thing is, you can still build a very high quality house with only a small amount of all the available tools.

Some tools should be owned by every homeowner. But some tools are expensive, large, or would rarely be used, making their purchase harder to justify. While those tools can be rented, other options should be considered because of the length of time it can take to build a tiny house.

RENTING

Tools are often rented at a daily rate between five and ten percent of the cost to purchase new. If you rent a tool for over twenty days, it will cost as much money or more than purchasing. This pricing makes sense for items that are rarely used or used for a very short time. An example would be a pressure washer used by a homeowner. A pressure washer is expensive, it requires maintenance, and may only be used for one day a year. In this case, paying 10% of its cost to rent it is reasonable. For most tools used for the duration of a construction project, this pricing does not make sense.

BUYING NEW

If you purchase tools new for your project, they can often be sold for at least fifty percent of the original cost. In this scenario, the tools can be used longer and would cost only half as much as renting for a short length of time.

BUYING USED

If the required tools can be found in relatively good condition, an even better option is to buy them used. A tool that is purchased used and only used for a few months adds little additional wear and tear. These tools can be sold for nearly the same price that was originally paid, effectively allowing the use of the tool for free.

I was able to do this a few years ago with a refrigerator, when I rented an apartment that did not come with one. At that time I had the option to either rent one from the apartment complex for $10 per month ($120/year) or supply my own. After research, I found that a new comparable model cost $200, while a used one between 1 and 2 years old cost about $100. The depreciation during the first year was significant, but the depreciation during the second year was negligible. I bought a one year old used refrigerator in near perfect condition for $100, then one year later sold it for $100. So I paid nothing for the use of the refrigerator and saved $120 in rental fees. This same method can be used for tools.

A good place to look into buying and selling used tools is Craigslist.org. Online marketplaces such as eBay are also good for purchasing smaller used tools, where shipping cost would not be prohibitive.

BUYING REFURBISHED

There are online retailers that are dedicated to selling refurbished tools. I have purchased several refurbished tools and have always had great success with them. They have arrived looking and working like they were brand new. When a manufacturer refurbishes a tool, they will usually modify the serial number to begin with an 'R' or an 'R' will be stamped into the casing to easily identify it as refurbished. If you buy any used tools, look for these markings and use them to negotiate a better price.

TOOL LIST

Below is a list of many of the tools that we use and a description of how we use them when building tiny houses.

THE ESSENTIALS

TAPE MEASURE - A good quality, sturdy 25 foot tape measure is a must. Like many of my hand tools, I try to buy a version that is brightly colored to make it easier to find. I have spent more time than I care to admit wandering around searching for my tape measure.

PENCIL - Carpenters have their own pencils for a reason, a thicker lead means less time sharpening and more strength while marking rough lumber.

CHALK LINE - This is used to mark a straight line over an extended length. It is essentially a string that is covered in chalk. The string is pulled tight before being pulled back and released to strike a surface where a mark is left behind. This tool is used most frequently while installing the exterior siding and flooring, to make sure everything stays straight. While using this tool, be careful that the chalk does not come in contact with any wood that will be visible when the house is complete. It can be very difficult to remove or cover up. Red chalk is the most difficult to remove, so blue chalk is recommended.

LEVEL - A level is used to verify that various surfaces are level, relative to the ground.

SQUARES - Squares come in various shapes and sizes. The most commonly used square is called a speed square because of its small size. Squares are used to assist in marking lines and making cuts that are 90 degrees.

HAMMER - This is one of the most commonly used tools. From hammering nails to knocking boards in place, a builder will not get far without one of these.

SCREWDRIVERS - Screwdrivers are used for miscellaneous tasks throughout the project, particularly while installing the electrical outlets, switches, and plates.

PLIERS - Pliers are not used that often except when installing electrical components. They are, however, an essential tool for any homeowner.

WIRE CUTTERS & STRIPPERS - Wire cutters are used while installing the electric lines. This is thick wire, so the larger the cutters the better. Strippers remove the sheathing from the electrical wires efficiently.

UTILITY KNIFE & BLADES - A utility knife is used to cut the housewrap, tar paper, and even to score the metal roof panels. Keep lots of blades handy, as they dull quickly.

STEEL SNIPS - Steel or tin snips are used to cut various metal components, primarily while installing the metal roof. They come in straight cutting, left cutting, and right cutting versions that designate in which direction they can easily cut angles.

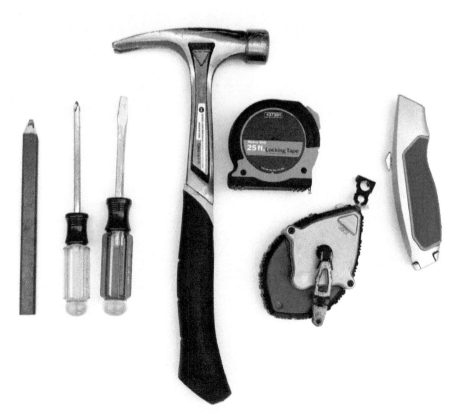

Carpenters Pencil, Screw Drivers, Hammer, Tape Measure, Chalk Line, and Utility Knife

Drill and Impact Driver

PVC Pipe Cutter and Hole Saw

PVC PIPE CUTTER - A pipe cutter is an inexpensive tool that creates quick and clean cuts of PVC and CPVC. A hand or miter saw can also be used as a substitute for this tool but it is not nearly as convenient and will leave behind burs that will need to be removed before the pipes are attached to any fittings.

HOLE SAW - A hole saw is a blade that attaches to a conventional drill. This tool is used to make holes in the subfloor for the drainage plumbing, as well as holes in the exterior walls for the bathroom vent and the plumbing exhaust.

DRILL - A drill is used to drill holes and to drive shorter screws. If the framing of the house is fastened with screws, this tool will get a lot of use when pre-drilling many of the framing boards

IMPACT DRIVER - Impact drivers are used to drive screws. While this tool may be confused for a drill, it works in a different way to produce a lot more torque than a drill. A standard drill might have a difficult time driving a 3½ inch screw without striping it; this tool will do it with ease.

CIRCULAR SAW - A circular saw is used to cut all of the sheathing and some 2x4's. Take extra care when using this tool, as with all saws.

Circular Saw and Extension Cord

Miter Saw

MITER SAW - A miter saw is essential to house framing. Behind the impact driver, it is the most commonly used power tool. Almost every single board used in your house, except for the sheathing, will have an edge cut with this saw. While a 10 inch saw will do 95% of the cuts, a 12 inch version with a blade that pivots in both directions is the best option.

GRINDER - A grinder is occasionally used on the metal roofing panels to cut sections where the snips or shears have a hard time reaching.

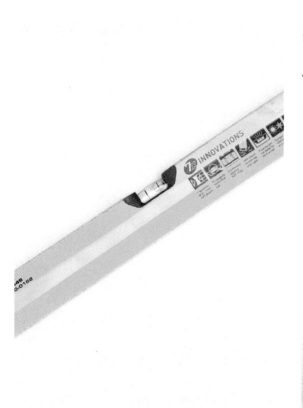

Level

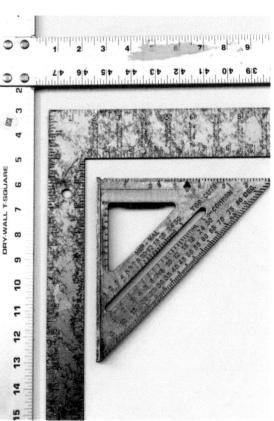

Drywall T Square, Carpenters Square, Speed Square

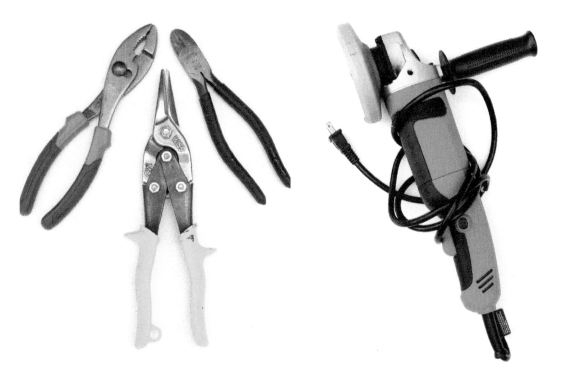

Pliers, Tin Snips, Wire Cutters

Angle Grinder

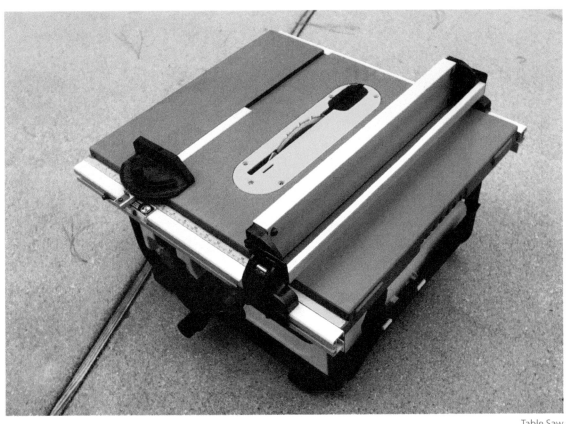

Table Saw

FLAT BAR - A flat bar is not used that often, but can be especially helpful with removing stubborn nails or adjusting sheathing and flooring. A straight claw hammer can often be used as a substitute.

TABLE SAW - A table saw is incredibly useful. It only made it to the 'almost essential' list because a circular saw can do a lot of what a table saw is used for, such as cutting boards lengthwise. A table saw does this job much more easily and with a lot more precision.

COMPRESSOR AND PNEUMATIC TOOLS - Since a house can be framed by screwing together the boards, a compressor and pneumatic tools are not required. They can, however, speed up progress considerably and I would not want to install interior plank paneling without a pneumatic brad nailer.

JIG SAW - A jig saw's narrow blade makes it unique because it can easily make curved cuts. This tool is not used often, but is really handy when cutting the sheathing and siding around the fenders. A hand coping saw can be used as a substitute.

POWER SHEARS - Power shears are used to make quick and clean cuts to a metal roof panel. A combination of snips and a grinder, or a utility knife can be used as a substitute.

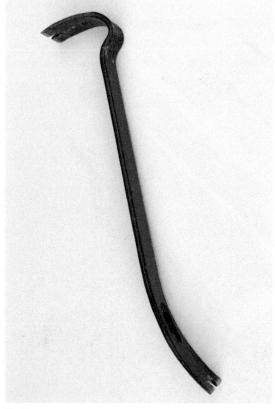

Flat Bar

Air Compressor

Jig Saw

Electric Power Shears

A comprehensive materials list is
essential

BUILDING MATERIALS

—

Having a comprehensive materials list can save a substantial amount of time and money. If a materials list was not provided with your plans or if you created your own design, you will want to take the time to come up with one before getting started. This can be done by mentally walking through each step of the construction process while examining the plans to count how much of each item or material will be required to complete each step. For instance, by examining the plans you can count the number of studs and sheets of plywood that will be needed. The problem with this approach is there are many items that could be overlooked if you do not have detailed knowledge of the build process. However, an incomplete list is still helpful as a starting point.

A material list allows you to purchase all of the required materials in as few trips to the store as possible. Not only will you spend less time going to and from the store, but you will also spend less time in the store. If a list is provided to one of the larger home improvement stores, they will 'pull' the order for you, saving you hours had you done it yourself. The down side to a store pulling an order is that the employees are not going to be as careful as you might be in their selection process, so you can expect to get a few items that may have been tossed aside by other shoppers. Even if you need to exchange a few items, this will require far less effort and time than gathering all the items yourself.

Many stores will provide special pricing for larger orders. In my experience, this discount has ranged from 90% on items like loose hardware down to 0% for items like tools. Typically, the overall discount will be about 10%. For a home requiring $12,000-16,000 in materials, this translates to significant savings. To get the discount, orders need to be placed at the office or counter reserved for professionals, sometimes called the Pro Desk. The orders must also be in excess of a minimum amount, usually around $2500. Because of this threshold it is important to maximize items and reduce the number of orders to benefit from the discounts available. If the store does not initially discount your order, you may need to ask them to send it to the 'bid room'. This is a way of telling them to send your compiled list to the store's corporate headquarters, where your discount will be calculated.

If you need to return some of the items in your order, reducing your order total below the discount threshold, the discount typically stays in place. So if necessary, you could pad your order by buying a few extra items to achieve the discount. I always recommend buying slightly more materials than you need anyways, as it can be frustrating getting some of your materials at 40% off, only to run out and have to buy more at full price in the store.

In order to avoid theft and weather damage with materials being stored outside, it is best to plan your orders in alignment with the stages of your tiny house project. You should divide your orders, while still making sure they are big enough to reach the price threshold for a discount. Place orders according to the sequence the materials will be used. By doing this, you will not have as many items laying around for extended periods of time.

If your materials are not stored inside, they will need to be covered by a tarp for protection. Be aware that moisture can build up under the tarp and result in mold growth. Mold can be easily removed from framing lumber, but will permanently stain and ruin expensive trim pieces. Trim is the most susceptible to damage, since it will be installed last and left outside the longest.

At the time of publishing this book, the big box retailers like Home Depot and Lowes also provide a special 10% discounts to veterans, with no minimum order required. Check with your local store for details on these and other discount options.

Below is a general list of materials that I often use when building tiny houses. If you are confused by some of the items, do not worry, they will be described in more detail in the later chapters. The quantities of each item will be determined by the design and size of the specific house being built, so that is not included here. Please consult your plan's material list, if provided, for those details. The items are grouped by the departments in which they are generally found in stores.

BUILDING MATERIALS

2X LUMBER - 2x (said "two-by") lumber has a 2 as the first dimension (e.g. 2x4, 2x6, etc.). Most of the framing is built with 2x4's. In conventional homes where trusses are not used, the rafters are usually 2x6 or larger. 2x4's are adequate for the small span of most

Building materials

tiny houses. Only where significant load from snow is expected would larger lumber need to be used.

The numbers used to describe lumber is not the actual size. See the chart below.

Nominal Size (inches)	Actual Size (inches)
2x4	1 ½ x 3 ½
2x6	1 ½ x 5 ½
2x8	1 ½ x 7 ¼
2x10	1 ½ x 9 ¼
2x12	1 ½ x 11 ¼
4x4	3 ½ x 3 ½

2x lumber also comes in varying lengths. Try to purchase lengths that will result in the least amount of waste. If your plans call for rafters that are 4 ½ feet long, it is better to order a single 10 foot length than two 8 foot lengths. This will not only reduce waste, but will save you money.

Trailer with Materials

A note on lumber in general: Always measure the lumber you are using and never trust the stated size. You will find that while some measurements are fairly consistently accurate, others can vary quite a bit. For instance, on multiple 2x4x8' pieces, they will all be very close to 1 ½ inches wide and 3 ½ inches high; but the lengths will vary by as much as a 1 inch. However, on multiple 2x4x96" pieces, their lengths will all be much more consistently accurate at 8 feet. You might be thinking that wood sold by the foot will not be as accurate to the stated size as wood sold by the inch. Plywood is sold by the foot, yet it is always very accurate. Confused yet? An easy solution is to measure everything. After a week or two you will learn what measurements to trust and which ones not to trust.

Another dimension of wood that you cannot always trust is its square-ness. Some wood is always very square (the corners are 90 degrees), like plywood, while other wood is notoriously not square, like siding. If you have wood that needs to be square, it is usually a good idea to run both ends through the saw. Again, a little experience will go a long way in knowing which boards will require this.

PLYWOOD – We strongly recommend plywood instead of oriented strand board (OSB). While plywood costs more, it is also more rigid, durable, and lighter than OSB. Additionally, when OSB gets wet it will swell up as it absorbs the water, but never returns to its original shape.

Plywood comes in 4 foot by 8 foot sheets of varying thicknesses and ply count. You will likely need two different thicknesses for your tiny house: 15⁄32 inch (often called ½ inch) for the walls and roof, and 23⁄32 inch (often called ¾ inch) tongue and groove (T&G) for the subfloor.

Note that tongue and groove sheets of plywood are 4 feet wide, but this includes the tongue that will overlap with any adjacent board's grooves, so two sheets together will not equal a full 8 feet.

EXTERIOR SIDING AND TRIM - Cedar is light and holds up well against the weather, making it a great exterior siding for tiny houses. This is a product that is often heavily discounted at the pro desk, so if you plan to finish the exterior of the house before the store's return period has elapsed, I suggest buying extra.

INTERIOR SIDING AND TRIM - Either slat paneling or sheet paneling is used for the interior siding because it is both lightweight and relatively durable. The types of trim include cove, casing, corner, and stop.

LOFT DECKING - For the loft decking, inexpensive plywood can be used. A more attractive alternative is 1x6 pine tongue and groove boards.

TAR PAPER - 30 lb. tar paper is used for under the metal roof. 15 lb. tar paper can be used as an option for underlayment under wood flooring.

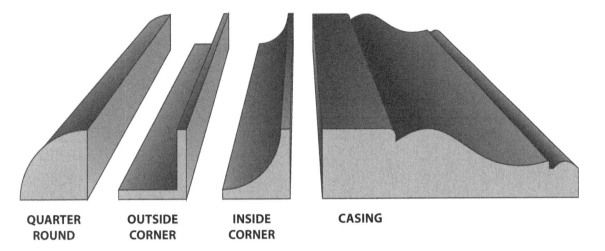

QUARTER ROUND **OUTSIDE CORNER** **INSIDE CORNER** **CASING**

Trim Profiles

WATER & ICE SHIELD – This product serves a similar function as tar paper, except it is more durable and backed by adhesive. Its higher cost usually reserves it for roof valleys on conventional homes, but on tiny homes it can be used to cover the entire roof. If it is to be used under a metal roof, be sure to get the variant that is made for high heat.

STRAPPING - Strapping are metal pieces used to reinforce specific connections in a house.

FLASHING - Flashing is an impervious material used to prevent water from coming in contact with wood or from entering a wall cavity. There are various pieces of flashing that are used, including 'Z' flashing that is used above and below the window and door trim, 3 inch roof edge flashing that is used to cover the fenders, and roll valley flashing that is used to protect the underside of the subfloor.

The flashing that is used as part of the roof is covered in the roofing section.

NAIL PLATES - Nail plates are attached to the studs wherever wires or plumbing go through the studs. These prevent the wires or pipes from accidentally being pierced by a nail or brad.

MILLWORK

FRONT DOOR - Front doors are generally made out of wood, steel, or fiberglass. Be wary of plans that call for a narrower, custom sized front door because this will likely be the only way to get furniture and appliances in and out of your house. You may not want to be stuck with the same refrigerator for the next 25+ years.

POCKET DOOR AND FRAME - A pocket door is built inside the wall cavity so that it can be opened and closed without getting in the way. The frames come either pre-assembled or in a kit. If the door is to be stained be sure to select one made from solid wood.

WINDOWS - Windows can make up a very large portion of the material cost of your house (up to 20 percent). They come constructed of different materials ranging from less expensive vinyl to premium aluminum clad wood. Aluminum clad wood windows have wood on the interior that can be stained but still have an extremely durable exterior that comes in a variety of colors. Vinyl windows are less expensive but not nearly as durable. They are also not available in all areas and generally only come in white.

SHIMS - Shims are tapered pieces of wood that are used while installing the windows and doors.

ELECTRICAL

LIGHTING - The lighting that is needed for the house will likely include a bathroom fan/light combo, a light over the shower, lighting for under the loft (recessed can lights), lighting for any open areas (ceiling fan/light combo), and any exterior lighting.

OUTLET BOXES - Outlet boxes are nailed directly to the studs and hold the switches and outlets. These come in different sizes depending on the number of outlets and switches (called 'gang') they will contain (e.g. 1 gang, 2 gang, etc.). These are also divided into 'new work' and 'old work'. For a new house you will choose 'new work'.

OUTLETS AND SWITCHES - Outlets and switches can normally be purchased in 'contractor packs' (10 per pack) to get a discount. Special switches (called 3-way switches) are required if you plan to operate a single light from two different switches.

SWITCH PLATES - Switch plates are installed on the outlet boxes to cover the switches and outlets.

WIRE - Residential sheathed wiring is run through the walls and provides power to the lights and outlets.

WIRE STAPLES - Wire staples are heavy duty staples that are installed with a hammer and hold the interior wiring against the studs. These are often used on the wires right before they enter an outlet box, in order to keep them in place.

ELECTRICAL PANEL AND BREAKERS - The electrical panel or breaker box will hold the breakers for the house's electrical system.

CABLE, NETWORK AND TELEPHONE WIRES - Any additional wires that you would like in the house will need to be purchased as well. It is best to install these cables away from the electrical wires as they can cause interference.

FLOORING

HARDWOOD FLOORING - Hardwood flooring is typically used throughout the house, with the exception of the bathroom.

VINYL TILES - Vinyl tiles can be attractive, lightweight, easy to install, and durable. These are a great option for the bathroom.

VINYL ADHESIVE - Vinyl tiles usually come with an adhesive pre-applied to the back of them, but an additional layer is recommended.

FLOORING TRANSITION PIECE - The flooring transition piece covers the transition between different flooring types and thicknesses.

HARDWARE

EXTERIOR SCREWS - Exterior screws are used to attach the sheathing and are an alternative to using nails for the framing.

WAFER HEAD SCREWS - Wafer head screws have a low profile head. These are used to attach the strapping to the framing, since they are less likely to interfere with the interior siding.

BOLTS, NUTS, AND WASHERS - Bolts are used to attach the subfloor to the trailer.

EXTERIOR NAILS - Spiral shank exterior nails are used to attach the exterior siding.

PLASTICAP OR SIMPLEX NAILS - Plasticap or Simplex nails have a large, flat plastic head. These are used to attach thin, less durable items like tar paper to the roof, or in some cases the housewrap.

STAPLES - Standard T50 staples are used to attach the metal valley flashing to the trailer before the subfloor is constructed. These do not have to be that strong since the subfloor will keep the flashing in place.

BRADS - Brads are small nails that have a very small head. They are often used on finishing products, where a visible nail head is undesired, like the interior siding.

DOOR HARDWARE- Door hardware is required for both the front door and the interior door.

INSULATION

EXTRUDED POLYSTYRENE (STYROFOAM) BOARDS - Styrofoam boards are used to insulate the subfloor and possibly the wall and roof cavities.

SPRAY FOAM - Several cans of aerosol spray foam will be needed to fill the cracks around the fenders, windows, and doors. This product comes in several varieties, based on how much it will expand. Minimally expansive foam is used around the windows and doors to avoid bending the jambs, preventing the windows or doors from operating correctly. Significantly more of these canisters will be needed if Styrofoam boards will be used for the interior insulation. In this case, I recommend finding a version, like that made by Hilti, which includes a rigid dispenser that will allow for one handed operation.

HOUSEWRAP AND TAPE - Housewrap creates both a moisture and air barrier around a house. Not all housewraps are created equal. I strongly recommend DuPont™ Tyvek®. A special tape is also required to seal any seams in the housewrap.

PAINT

PAINTS AND STAINS - Paints or stains are used for the exterior siding, the interior siding, and the front door.

CONSTRUCTION ADHESIVE - Construction adhesive, sometime referred to by the brand name Liquid Nails, is used when installing the sheathing and interior siding.

PLUMBING

FIBERGLASS SHOWER - Several different materials can be used to make a shower stall. A popular option is a one piece, 36 inch fiberglass shower. Due to its size, the shower needs to be put in the house before all of the exterior walls are stood up. These showers also come in kits that can be installed later, but they are generally flimsy and more difficult for the homeowner to clean because of the seams.

CPVC PIPES, FITTINGS AND PRIMER/CEMENT - CPVC pipes and fittings are used for the plumbing supply lines.

PVC PIPES, FITTINGS AND PRIMER/CEMENT - PVC pipes and fittings are used for the plumbing drain lines.

KITCHEN SINK AND FAUCET - A narrow, single well sink should be selected to preserve counter space. If a larger sink is used, a cutting board can be used to cover it up to temporarily regain some counter space.

SHOWER ASSEMBLY - The shower assembly includes the shower head and the controls for the water pressure and temperature.

TOILET - The available utilities at the location where you plan to park your house will determine the type of toilet you need. Popular options include a standard flushable toilet and a composting toilet.

WATER HEATER - For a tiny house, a tankless water heater is optimal. They have a smaller size and provide endless hot water. While there are several inexpensive options available, if your house will be located in an area that will often have cold weather, be sure to choose a heater that is or can be insulated.

DRYER VENT - A dryer vent is used for the bathroom fan and also to cover up the air inlet for the plumbing.

HANGER TAPE ROLL - Hanger tape is galvanized strips of metal used to add support to hanging pipes.

Great Stuff Brand Spray Foam

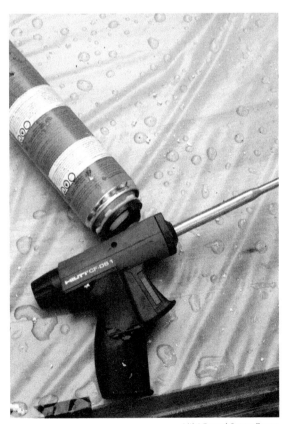
Hilti Brand Spray Foam

TRAILER SUPPLIER

TRAILER - The trailer is the foundation of a house on wheels.

JACK STANDS - Jack stands are used to level the house during construction and to keep the house stable while it is lived in.

METAL ROOF

The metal roofing will likely need to be custom ordered, as retailers generally only carry basic parts that are more suitable for sheds. Metal roofing is made up of several different components. The complexity of your roofs design will dictate how many components you will need to order.

The diagram to the right illustrates the most commonly used metal roofing components.

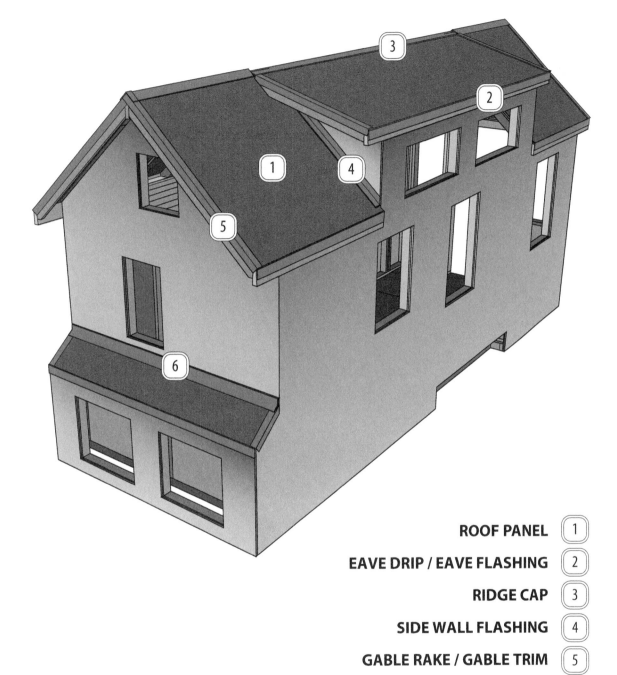

ROOF PANEL ①

EAVE DRIP / EAVE FLASHING ②

RIDGE CAP ③

SIDE WALL FLASHING ④

GABLE RAKE / GABLE TRIM ⑤

END WALL FLASHING ⑥

Metal Roofing Components

Getting the right trailer for your design is important, as it is the foundation of your house

TRAILER

TRAILER TYPES

Trailers are generally categorized by what they will be used to haul. While there are no industry standards that define these categories, you will often find trailers that are best suited for tiny houses referred to as either equipment trailers or car haulers. There are also trailers designed specifically for tiny houses which come ready to be built on.

UTILITY TRAILER

Utility trailers are lighter duty trailers typically used to move landscaping equipment or all-terrain vehicles. They usually feature side rails and a large ramp at the back to facilitate loading and unloading equipment. Because of their smaller axles and thinner steel, this trailer type is not suitable for a tiny house.

EQUIPMENT TRAILER

Equipment trailers are heavy duty trailers typically used to move heavy equipment like tractors. Variants of these designed specifically to move cars are called Car Haulers. Because of their heavy duty axles and thicker steel, these trailers are often used as tiny house foundations.

Utility Trailer

Equipment Trailer

DECKOVER TRAILER

Deckover trailers are very similar to equipment trailers. The main difference is that with deckover trailers, the deck is raised above the fenders allowing the top surface to be completely flat. While a flat surface may seem appealing for the purpose of building a house, the tradeoff is that this reduces the interior height of your home, since there is an overall height limit of 13 ½ feet. The height difference between a standard equipment trailer and a deckover is typically around one foot. Therefore, this trailer type would only be suitable for a house without a sleeping loft. Also be aware that raising the deck height will also add 2 additional steps leading to the house.

GOOSENECK TRAILER

A gooseneck trailer is also very similar to an equipment trailer. The main difference between the two is the hitch. A gooseneck trailer has a much larger hitch assembly that is raised up and attaches to a special receiver installed inside the bed of a truck. If you are familiar with RV's, a 'fifth wheel' trailer has a gooseneck. Gooseneck trailers require a larger tow vehicle, with an expensive receiver installed. The tradeoff is that they can carry more weight and can be easier to tow. Gooseneck trailers are not commonly used for tiny houses since they have a different aesthetic to them that is typically associated with RV's.

Deckover Trailer

Gooseneck Trailer

Tiny House Trailer

TINY HOUSE TRAILER

A tiny house trailer is a trailer specifically designed to be built on. It has features that will make that job easier and will not require the preparation needed when using an equipment trailer.

ACQUIRING A TRAILER

When shopping for a trailer, it is helpful to know the names of the various parts. See the diagram to the right for the location of the different components.

A trailer can be purchased either new or used. For a used trailer, I suggest checking Craigslist and local trailer dealers. In my experience, however, I have not found there to be a substantial price break in used trailers unless they required significant restoration work. Considering that the trailer is the foundation of the house, it is important that there are no issues with it. These potential drawbacks, coupled with the customizations that can be made to a new trailer, I recommend buying new.

TRAILER FEATURES

Trailer manufacturers add features to their trailers specifically to aid in the job that they are intended to perform. For example, if a trailer is made to transport a car, it needs to

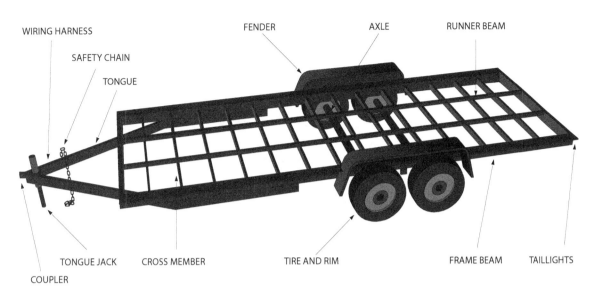

WIRING HARNESS

SAFETY CHAIN

TONGUE

FENDER

AXLE

RUNNER BEAM

COUPLER

TONGUE JACK

CROSS MEMBER

TIRE AND RIM

FRAME BEAM

TAILLIGHTS

Trailer Diagram

Trailer Tongue

have 'tie down' points in order to secure the vehicle to the trailer. If you are buying a trailer that is not specifically made for a tiny house, it may have features that make building on it difficult. If you are having the trailer custom manufactured, you can request which features you want. Features that make a trailer better suited to be used as a tiny house's foundation are:

NO DOVETAIL

A dovetail, sometimes called a beavertail, is a slight decline at the end of a trailer's deck. This feature facilitates loading equipment by reducing the incline angle of the trailer's ramps. Since the trailer's deck supports the subfloor of the house, a slant in the deck surface is undesirable. If you are required to use a trailer that has a dovetail, wooden supports can be constructed to fill in the gap created by the dovetail's angle. Since a tiny house's siding does not generally extend below the trailer's deck, this will be visible when the house is complete and may look unattractive.

NO TOP RAIL AND UPRIGHTS (SIDEWALLS)

Since the subfloor of the home will either come to the edge or extend beyond the width of the trailer deck, sidewalls should be avoided. If necessary, steel sidewalls can be removed by using a metal grinder/cutter.

NO FRONT GUARD

A front guard helps to prevent equipment from pushing past the front of a trailer into the tow vehicle. This is undesirable but can be built around if necessary.

Note that the trailer in some of the pictures has a front guard.

HEAVY DUTY AXLES (5000 - 7000 lbs. EACH)

On shorter trailers, less heavy duty axles may be sufficient. However, on longer trailers larger axles are recommended.

NO RAMPS

Ramps are generally installed on equipment trailers for the loading and unloading of equipment. They are designed to be raised and lowered and are attached to the trailer with a large pin. While the ramps themselves will not be welded to the trailer, the pin may be. If purchasing a trailer from a dealer, have them remove the ramps. Otherwise, you can attempt to hammer out the pin, but be sure not to damage the trailer. If the pin does not slide out, you will need to take the trailer to a metal shop to have the ramps removed.

Tiny House Trailer

Equipment Trailer

To assist in securing the house's foundation to the trailer and to reduce the amount that the subfloor overhangs the trailer's edge, a piece of angle iron can be welded to the sides of the trailer so that it is even with the top of the trailer deck. Individually attached brackets can also be substituted for this and are covered in the chapter on strapping.

LEVELING THE TRAILER

The trailer needs to be leveled before any construction can begin. To assist with this I recommend using an SUV Jack and adjustable 6 ton SUV jack stands as supports, which can also be used to stabilize the house when it is completed. The adjustment granularity on the jack stands is not very fine, so boards of various thicknesses may also need to be placed under the stands to make minor adjustments. Regardless of the type of supports selected, be sure that they are strong enough to not only support the weight of your trailer, but also the completed house.

Try to build on a relatively level surface, as this will greatly simplify this process.

To begin, place supports under each of the four corners of the trailer. If the house is to be built on grass or soil, place the supports on cement pavers to reduce settling. Be sure the pavers are sized so that the entire base of the support is on the paver. Even with the use of pavers, some settling may occur over time, requiring the house to be re-leveled.

Identify and start at the highest corner of the trailer and use a jack to bring the remaining corners up to that level, adjusting the supports as you go. The jack that is built into the trailer can also be used to adjust the height at the front of the trailer. When checking if a side is level, only place the level on the metal framing of the trailer to reduce inaccurate measurements from irregularities in the trailer's deck boards.

Expect this process to require adjustments on each side or corner multiple times. This can be time consuming, but it is important to take the time to complete it correctly before moving on.

Trailer with Jack and Support

Trailer Being Leveled

Trailer Being Leveled

The subfloor of a tiny house is a
composite of the metal trailer and the
wood constructed subfloor

SUBFLOOR
FRAMING &
INSULATION

—

0

In a conventional shed or house, the subfloor framing is generally constructed of 2x6 or larger lumber. In a tiny house, 2x4 lumber can be used because the subfloor will be resting and getting support from the metal framing of the trailer. While thicker lumber can still be used, which is usually only considered to create a larger cavity for additional insulation, the thinner lumber allows for more room in the interior of the house since the overall height of the house is limited.

PREPARING THE TRAILER

If you are using a trailer other than a tiny house trailer, the first step is to prepare the trailer. Depending on your particular trailer and its features, the preparation work required will vary. Consult the section on trailers to identify any parts that may need to be removed. If your trailer has a wooden deck, every other board from the deck will need to be removed. This is done to reduce the total weight of the house. Additionally, these boards will not bear much weight compared to the subfloor framing boards that run perpendicular to the decking. Before removing the boards, verify that none of the trailer's electrical wiring is attached to any of them. These trailer boards should not be used in the construction of the house, since they are pressure treated. Pressure treated wood is treated with chemical preservatives to make it more resistant to insects and decay. It is not used in tiny home building because it is heavy, it outgasses, and it contains metals that can react with the metal in your fasteners. When dissimilar metals come in contact with each other a galvanic reaction occurs which can rapidly corrode fasteners.

Trailer with Boards Removed

ADDING FLASHING

Next, flashing is installed over the entire trailer deck surface to protect the underside of the house, specifically the exposed insulation that will be added later. There are two materials that can be used as flashing.

ROLLED GALVANIZED FLASHING

If you are using a standard equipment trailer with a wood deck, rolled galvanized flashing is the easiest to install. A hand stapler (using T50 staples) or roofing nails can be used to attach the flashing to the wood. The flashing should overlap by at least 4 inches.

If you are using a tiny house trailer, then there is no wood on the deck to attach the flashing to. To get around this, you can attach the rolled flashing to the underside of the subfloor before installing it. An even easier option is to use metal roofing.

INVERTED METAL ROOFING

Metal roofing panels are easier to work with since they are rigid. The panels selected should have a profile that includes large flat sections. This will make them easier to line up so that they do not interfere with the trailer frame. The roofing material must be installed upside down, with the raised portions facing down. This way, they will not

Trailer with Flashing Installed

interfere with the subfloor installation.

Next, if your trailer has curved fenders, create a template that will be used later when cutting the sheathing and siding. This template can be laid on the sheathing to replicate that curve when it needs to be cut in a later step. This is accomplished by holding a board up behind the fender and drawing a line that matches the curve of the fender. Use a jig saw to cut along the line and create the template.

Before beginning any building, verify the measurements from the plans with the physical measurements of the trailer. This is particularly important for the cutouts for the fenders, if applicable, since the fenders may be in a different location than those in the plans.

The subfloor can be divided into three sections: in front of the fenders, behind the fenders, and between the fenders. Each section can be built and moved separately for convenience.

When constructing the subfloor or wall framing, it is easiest to mark the kick plate and the top plate at the same time. This ensures that all the boards will be properly aligned.

Tiny House Trailer with Metal Roof Flashing Installed

Top Plate and Kick Plate Marked

Top Plate Pre-Drilled

Note: Before constructing the subfloor, ensure that the joists are not lined up with the metal support beams (cross members) under the trailer's decking. If they are, the subfloor framing will need to be shifted so that the subfloor can be properly attached to the trailer. Later on, holes will be drilled through the subfloor and into the decking. The metal supports should not be in the way.

FRAMING FASTENERS

Either screws or specialized nails can be used to join the pieces of wood for the framing.

SCREWS

If screws are to be used, the first board that the screw is to pass through will need to be pre-drilled; otherwise, the boards will not bind together when the screws are tightened. If the boards are not pre-drilled, you also risk splitting them, especially when driving screws near the ends. A 3½ inch exterior screw should be used. Driving a 3½ inch screw requires a large amount of torque and a standard power screwdriver will likely only strip the head of the screw. To avoid this problem, use an impact driver. These are very effective, but also louder than a standard power screwdriver.

If nails are to be used, choose either ring shank or spiral shank nails that are more difficult to remove. Ring shank nails have small rings or ridges around them. Spiral shank nails, sometimes called screw nails, are twisted so that they rotate as they are driven. Both of these nails may also come with an adhesive on them to further hinder their removal. Nails are easier to pull out than screws, but the wall will receive additional support from the sheathing that will be screwed and glued to the framing. The primary benefit of using nails is that a pneumatic framing nail gun can be used to drive the nails, significantly speeding up the build process.

ATTACHING THE SUBFLOOR TO THE TRAILER

On the top surface of the trailer deck flashing, use a chalk line to mark the center of the beams or boards that run the length of the trailer. Move the subfloor sections into place. Where the joists and chalk lines intersect, use a one inch paddle bit to drill down into the top surface of the joist, deep enough so that the head of a hex bolt and washer will not protrude above the surface when installed. Then, in the center of that hole, use a 3/8 inch drill bit to drill down through the remaining section of the joist and through the beam or board below. Repeat this for each possible attachment point.

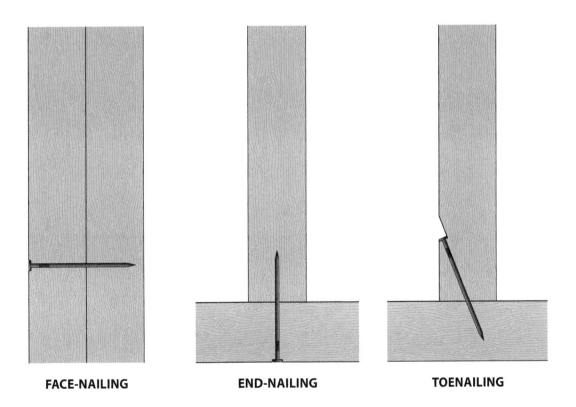

FACE-NAILING   **END-NAILING**   **TOENAILING**

Nailing Methods

Subfloor Framing

Bolt Through Joist

Bolt and Nut Below Trailer

It can be helpful to secure the subfloor sections to the trailer at just a few points before drilling all of the holes. A small shift in a subfloor section during drilling can result in a lot of wasted work, as the holes may no longer line up with the trailer beams. To attach the subfloor sections to the trailer, use 3/8 inch galvanized hex head bolts sized long enough to penetrate through the subfloor and trailer, leaving approximately 1 inch of bolt sticking out. Add washers to each end of the bolt and secure it with a 3/8 galvanized nut.

Having the small amount of excess bolt will be helpful if the nuts ever need to be re-tightened. Pliers can be used to grab that extra section of the bolt to prevent it from spinning, while tightening the nut. If the bolt is installed in the opposite orientation, with the nut on top, there will be no way to re-tighten the bolts later on.

SUBFLOOR INSULATION

The next step is to add the subfloor insulation. Since there are no wires or plumbing in the subfloor, it is best to fill as much of this cavity as possible. Use foam panels, in a combination of sizes, to achieve the 3 ½ inch depth needed. Cut the panels so they are approximately 1 inch smaller than the cavity that needs to be filled. Center the panels in the cavity and use a spray foam insulation product to fill the ½ inch gap around the edges.

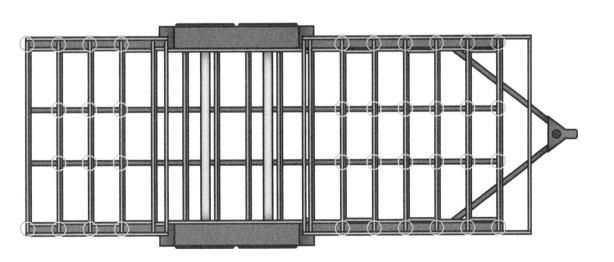

● **ATTACHMENT AT FLANGE** ○ **ATTACHMENT AT RUNNER**

Attachment Points
___

Subfloor Framing Attached to Trailer

Subfloor Framing Completed

Subfloor Insulation

Subfloor Insulation Being Installed

The subfloor sheathing will support
the weight of all the items and
occupants in the home

SUBFLOOR
SHEATHING
—

The subfloor sheathing is 23/32 inch tongue and groove plywood. The sheets are placed with the grain perpendicular to the floor joists. They are then attached to the subfloor framing with a construction adhesive and 2 inch exterior screws. You should apply the construction adhesive directly to the joist, with care being taken to keep the adhesive from coming in contact with the foam insulation. Some brands will actually dissolve it. The screws should be 12 inches apart on the inner part of the sheathing and 6 inches apart along the edges.

Avoid placing screws along any edge that will have another piece of sheathing placed against it until both are in place. Screws can sometimes pinch or splinter the sheathing, making it difficult to properly place another board tight up against it. When the screws are installed along edges shared by two pieces of sheathing, stagger them so that the screws on one piece of sheathing do not line up with the screws on the other.

You can use a sledge hammer to tighten two pieces of sheathing together along the tongue and groove side. If a sledge hammer is used, do not strike the sheathing directly. Instead, hold up another piece of board against the sheathing and strike it. This will reduce the chance that you accidently damage the sheathing.

As mentioned earlier, while individual sheets of T&G plywood are 4 feet by 8 feet, this measurement includes both the tongue and the groove. This is important because the tongue on one piece will overlap the groove on the next, resulting in a combined size of less than 8 feet.

Sheets of plywood are very square when they come from the factory, so the edges of the plywood should line up very closely with the edges of the subfloor framing. If this is not the case, double check the squareness of the framing and do not assume it is because of an imperfection in the wood.

Subfloor Sheathing Half Installed

Much of what was learned in framing
the subfloor also applies to the walls

WALL FRAMING
—

1

The wall framing is essentially the same as the subfloor framing. The big difference is that you will need to account for window and door openings in the walls.

Begin by cutting all of the framing boards (e.g. studs, headers, etc.) needed to complete one section of the wall. Mark both the top plate and the kick plate with the location of each stud at the same time. This will ensure that the studs are vertical when attached. Before the wall is erected, it is also a good idea to mark the top of the top plate for the rafters, as doing this later will require a ladder.

The lines for the stud locations are marked using a speed square, which is sized perfectly to cross over two 2x4's.

Each wall section is constructed laying down and then stood up after it is assembled.

Attach temporary stops along the edge of the subfloor to prevent the wall from sliding off the floor. Stops will also ensure that the wall is properly aligned with the edge once it is erected.

Once the wall is in its proper location, use 3 ½ inch screws every twelve inches to secure the kick plate. Consider using temporary bracing to ensure that erected walls do not fall over before they are properly secured to an adjoining wall. If an adjoining wall is going to be stood up immediately, and it is not windy outside, supports may not be necessary.

| Notes | Top and Kick Plate Marked |

Next, work on an adjoining wall so that it can be used to provide stability to the first wall section. At this stage, the walls should only be attached to each other with a minimal amount of screws. They may need to be removed later when the walls are squared and leveled.

HEADERS

Some wall sections may require a header. A header is a section of the wall frame designed to bear weight and bridge an opening. Headers are usually found over a door or window wide enough to intersect at least one load-bearing stud.

A header is composed of two 2x (e.g. 2x4, 2x6, etc.) pieces of lumber with a ½ inch piece of filler sandwiched in the center, all cut to the proper length and nailed or screwed together. The filler can be either plywood or foam insulation. The combined width of these materials is 3 ½ inches, making it the same width as the wall. When installing a header, be sure the orientation is such that the filler edge is visible from the top and bottom.

As with the other sections of the wall framing, if the design plans call for any higher level walls, like those found on a dormer, I recommend constructing these sections on the ground and then lifting them into place.

Wall Partially Completed

Temporary Bracing at Subfloor Edge

Wall Nailed and Screwed In Place

Wall Stood Up and Braced

Adjoining Walls Stood Up

Header Components

Header Assembled

Upper Wall Section Parts

Upper Wall Section Assembled

Wall sheathing is attached to the
outside of the studs, providing
support to the framing and a base for
the siding

WALL SHEATHING
—

The wall sheathing material is 15/32 inch plywood. It is attached with construction adhesive and 2 inch screws. Exterior wall sheathing should be installed vertically. This orientation provides the most strength, since there are no unsupported or unscrewed seams.

To attach the sheathing, begin in a corner of the house and cut a sheet to the correct size, with any openings required. Mark the stud locations on the sheathing to assist with screw placement later. Add construction adhesive to the studs, top plate and kick plate in the area where the sheet is to be attached. Attach a few temporary support pieces under the subfloor overhang to rest the plywood on. This will save your back and ensures that the sheathing is even with the bottom of the subfloor once stood up.

A sheet of sheathing can be expected to be very square; all four corners are 90 degrees. You can assume the wall is square if the sheet of sheathing is even with both the bottom of the subfloor and outside edge of the wall.

Verify this with a level and make any fine adjustments, as necessary, before screwing it in place. You may need to remove screws securing the wall to the adjacent walls, so that the wall can be easily moved and made square.

On the outside edges of the sheathing, the screws should be approximately 6 inches apart. On the inside studs, screws should be 12 inches apart. Offset or stagger the screws

Walls Sheathed

End Wall Sheathed

Temporary Bracing Under Subfloor

along the seams of adjoining pieces of sheathing.

FIXING WARPED PLYWOOD

If the sheathing of the house gets wet before you have a chance to protect it with the housewrap, it may warp. If the wood warps and protrudes into the wall cavity, it will not interfere with the installation of the siding; however, it will reduce the amount of insulation that can be used in that section of the wall. In this case, correcting the warped material is optional. On the other hand, if the wood warps out from the wall, then fixing it will be required. At this stage, it will likely be easier to repair the wall than to replace components, since the sheathing is glued into place.

To repair warps, set the blade depth of a circular saw to approximately half the thickness of the sheathing. Cut several expansion slots into the sheathing, on the side where the sheathing is expanding. On the inside, place a 2x4 against the sheathing and screw into it from the outside. As the 2x4 is screwed into, it will be pulled tight against the sheathing and will straighten it.

Warped Sheathing

Warped Sheathing

Set Blade Depth

Expansion Cuts

Wall Braced on the Inside

Wall Screwed from the Outside

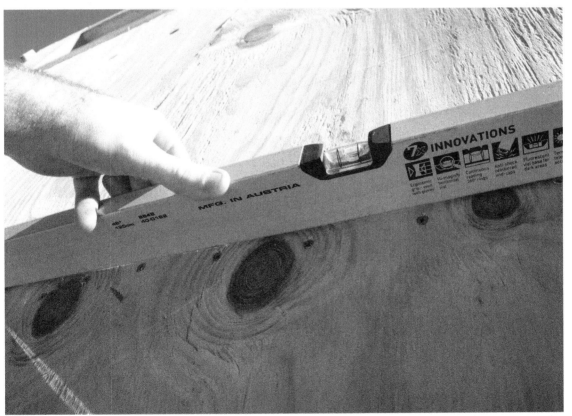

Wall Sheathing No Longer Warped

Framing the roof is more challenging
than framing the walls, since much
of the work needs to be done on a
ladder

ROOF FRAMING

Before beginning work on the roof, the walls should be braced to ensure that they do not push out while installing the rafters. Brace the walls by screwing a small section of 2x4 to the floor about 2 feet from the wall and lined up with a stud. Next, attach a longer piece of 2x4 to the floor piece and then to the corresponding stud. As the bracing is attached to the stud the wall should be made level and straight. This should be repeated every 4 to 6 feet along the length of the house. The walls can also be braced to each other by nailing a board across the top plates of the walls, ensuring they do not spread apart. Be sure not to place this support where the rafters will be located.

The next step is to cut the rafters. Before cutting all of the rafters for the entire roof, double check your cuts and angles on just a few pieces. A quick check can be accomplished by holding two rafters in place, either on the top plate or someplace similar like the sill plate of two adjacent windows (if your design has two windows of equal height across from each other).

Mark both sides of the ridge board for the rafter locations. The ridge board should then be lifted up into place and the rafters nailed to both it and the top plate of the walls. While nailing the rafters into place, it will be necessary to either have multiple people hold the ridge board or to use a temporary support.

In order to accommodate the length of most tiny houses, the ridge board will likely need to be two pieces of wood that will be joined together by a third, shorter, piece called a

Rafters Being Installed

Rafters Nailed

scab. If a solid piece can be used, it is preferred.

Assuming that the walls are level and evenly spaced across the entire length of the trailer, and that the rafters are all cut to the same size, each rafter should meet up with the ridge board at the same place. Since the ridge board is such a long piece of wood, it will likely not be perfectly straight. You should move it into the correct position, as needed, to consistently line up with each rafter, instead of cutting the rafters to fit. The point is to not let a curved piece of wood determine your ridge.

Note: A technique often used by contractors, especially when working with angled pieces of wood like those found around the roof, is scribing. When there is a piece of wood that needs to have an angled cut, it can often be difficult to replicate that angle in order to cut it to fit. Scribing is when you hold a piece of wood as close as possible to where it needs to go, and then you trace the angle or pattern that you need to cut into the wood. A special tool is available to aid with scribing, called a scribing or marking gauge, but for most of my cuts I am able to use my hand to get good enough results.

An internet search for "scribe technique" can provide additional details.

Temporary Support for Ridge Board

Rafters at Ridge Board

Metal pieces called strapping are used to reinforce specific connections in a house

STRAPPING

—

To ensure a secure connection between the various layers of a house's frame, strapping is added in specific locations.

Strapping comes in many different varieties for different applications, but can most easily be thought of as a metal strap that is screwed to two or more pieces of wood, reinforcing their linkage.

For strapping to be effective, the fasteners in the strap should be perpendicular to the shear line that is created by a force. For example, when wind hits the side of a house, in addition to applying direct pressure to that side, it also travels up the wall and puts upward pressure on the overhang. Since the rafter boards are nailed down into the top plate of the wall, the nails are aligned with the force of the wind. This makes the rafter susceptible to being pulled away from the wall. An example of this problem is older homes, that are not equipped with hurricane straps, losing their roofs in hurricanes. By adding a strap, the fasteners of the strap are perpendicular to this force and would need to be sheared for the rafter to be freed. Shearing a nail or screw is much more difficult than simply pulling it out.

To attach strapping, use either nails or 1¼ inch wafer head screws that are designed for straps. Wafer head screws have a low profile head and are less likely to interfere with other building materials, like the interior siding, later on.

Subfloor Bolted to Trailer

STRAPPING TYPES

Let us examine the various layers of the house's frame and their linkages to identify the need for strapping and the strapping types:

TRAILER'S METAL FRAME TO TRAILER'S WOOD DECKING
(NON-TINY HOUSE TRAILER ONLY)

The linkage between the wood deck and the metal frame in most trailers are heavy duty screws, added when the trailer is manufactured. The designed purpose of these screws is to keep the trailer decking in place while carrying a heavy load, like a piece of heavy equipment. These screws are not designed to sustain the upward force that can be exerted by the attached house while driving and rounding a corner or while encountering strong winds. Since there is not a very strong link in this direction, we cannot rely on just securing the house to the decking boards. While the house will be attached to the decking boards it will also need to be directly attached to the metal framing of the trailer. Additional screws or bolts can also be added between the decking and the metal framing to increase this linkage's strength.

If you are using a trailer specifically designed for a tiny house, then the wood is replaced with steel beams that are welded in place, requiring no additional support.

TRAILER'S WOOD DECKING TO HOUSE'S SUBFLOOR
(NON-TINY HOUSE TRAILER ONLY)

This linkage is made by approximately 30 to 50, 3/8 inch hex head bolts driven through the decking boards and subfloor framing. The exact number will depend on the length of your house.

TRAILER'S METAL FRAME TO HOUSE'S SUBFLOOR

If you are using a tiny house trailer, all of the connections between the subfloor and the trailer will be to the metal of the trailer.

If you are using a standard equipment trailer, there are several different methods to create this linkage. One option is to have the trailer manufacturer weld a piece of angle iron along the sides of the trailer, flush with the deck, to create a flange that can be screwed up through and into the subfloor framing. A less expensive, although not as strong, alternative is to use angle brackets that are bolted to the trailer framing to achieve the same goal. These brackets would only be attached to the trailer at the locations where the joists extend beyond the edges of the trailer.

Bracket Attaching Subfloor Framing to the Trailer

Brackets after Being Painted

Strapping for Upper Walls

Hurricane Strap

Strap Connecting Rafters

HOUSE'S SUBFLOOR TO HOUSE'S WALLS

The walls are secured to the subfloor using two methods. First, 3½ inch screws are driven through the kick plate into the subfloor at 12 inch intervals. Second, a much stronger linkage is created by the wall sheathing. The wall sheathing covers the entire wall and also extends down to overlap the edges of the subfloor. Since the sheathing is secured with 2 inch screws and construction adhesive, it creates a very strong linkage.

HOUSE'S WALLS TO UPPER HOUSE'S WALLS

Upper walls are any wall portions that sit on top of a lower wall's top plate. For instance, the walls found in dormers. If possible, the sheathing seams should not be lined up with the framing seams between these two layers, as the sheathing can be used to reinforce this linkage. Since it may not be possible to avoid lining up these seams, or the overlap may not be substantial, it is also recommended to add metal strapping across these layers.

HOUSE'S WALLS TO ROOF'S RAFTERS

A specialized type of strapping, called hurricane straps, are used for this linkage. These straps have a bend in them to facilitate attaching them to both the top plate of a wall and a rafter.

RAFTERS TO RIDGE BOARD

Strapping is suggested here as well. There are various options available, but a single 18 inch strap that wraps from one rafter, over the ridge board, to the other rafter should provide adequate support. This strap is intended to prevent the rafters from being pulled away from the ridge board. Collar ties can also be used, which are pieces of wood that run underneath the ridge board and connect the two rafters together.

The roof sheathing is installed very
similarly to the wall sheathing

ROOF SHEATHING

—

Like the wall sheathing, the roof sheathing material is 15/32 inch plywood and is attached with construction adhesive and 2 inch exterior screws. If the rafters are 24 inches apart, plywood clips, also known as H clips, should be placed between the horizontal seams of any two adjoining sheets and centered in the area between the rafters. The grains of the plywood sheets should be perpendicular to the rafters for strength, with the long side of the sheet parallel to the ground. Stagger the seams so that a lower sheet's seam does not line up with any adjoining upper sheet's seam on the same rafter.

Install temporary blocks along the bottom edge of the roof to prevent the plywood sheets from sliding off while you are putting them in place. These also ensure that the plywood will be even with the bottom edge of the roof. Blocks can be made of scrap material, including small pieces of plywood or 2x4s. They should only extend above the edge of the roof about one to two inches; otherwise, they may get in the way and make it more difficult to lift the sheathing into place. Cut a sheet of plywood to the size required and mark the rafter locations to assist with screw placement later. Apply construction adhesive on the rafters that the sheathing will lay on and put the sheet in place. If construction adhesive gets on any part of a rafter that will not get covered by sheathing in a relatively short period of time, be sure to clean it off before it hardens. Removing hardened adhesive is considerably more difficult than removing it when it is fresh and wet.

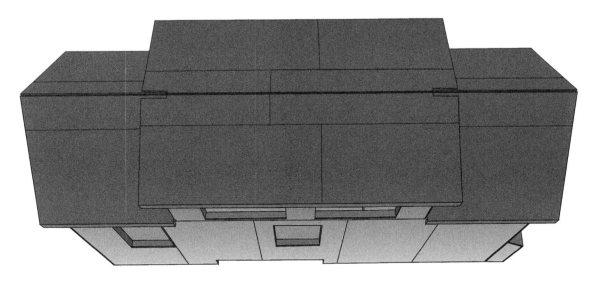

Sheathing Seams Offset

Roof Sheathing Ready to be Slid into Place

If working alone, consider using smaller pieces of sheathing. Smaller pieces will often result in more waste, but are also a lot easier and safer to move, especially while on a ladder.

A technique to get larger pieces of sheathing up on a roof is to use an extension ladder. The ladder is positioned to lean against the roof at a relatively gentle angle. The sheathing is centered at the bottom of the ladder and slid up as you walk up the ladder behind it. The sheathing is pushed up until it slides off the top of the ladder and on to the rafters. If this technique is used, care should be taken so that the base of the ladder does not lift off of the ground when the sheathing gets to the top, as this can result in the ladder sliding and falling.

Once the sheet is in the proper location, attach it with 2 inch exterior screws. On the outside edges, the screws should be approximately 6 inches apart. On the inside rafters, they should be 12 inches apart. Offset the screws along the seams of adjoining pieces of sheathing.

Roof Sheathing in Place

When a house is protected from the
weather, it is considered dried-in

DRYING-IN THE HOUSE

—

15

Once all the sheathing is complete, the first layer of weather protection is added. This is referred to as drying-in a house. When your house is dried-in, you can finally sit back and relax for a moment and not worry as much about the weather affecting your building materials and investment.

ROOF

There are two options for protecting the roof of your house, tar paper or water and ice shield.

TAR PAPER

Tar paper, sometimes called felt, is the traditional roof weatherproofing material. It has been used successfully for many years and is sold on a roll in different weights or thicknesses. For use on a roof, 30 lb. weight should be selected. It is attached to the roof with plasticap or simplex nails. The installation of the tar paper should begin on the lowest part of the roof, with the next highest section added afterwards. This ensures that the higher sections overlap the lower sections, reducing the chance of rain water getting under the paper. Each section of tar paper should overlap the lower section by at least 2 inches, but 6 inches is preferred.

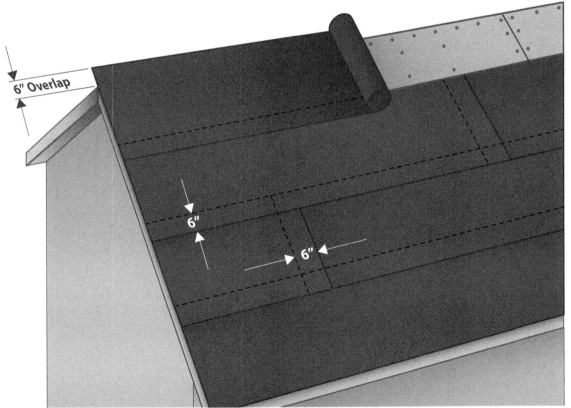

Tar Paper Overlap

Tar Paper Installed

WATER AND ICE SHIELD

Water and ice shield is used in much the same way as tar paper. The biggest difference is that it has an adhesive backing, so it sticks directly to the sheathing and does not require any fasteners. However, the adhesive backing makes it a little harder to install, since it has a tendency to stick to itself and is difficult to adjust. In traditional homes, because of its expense, water and ice shield is normally reserved for areas that are more prone to leaking like roof valleys. On tiny houses, because of their smaller size, it can be installed over the entire roof. While more expensive than tar paper, it is very effective and worth your consideration.

WALLS

To protect the exterior of the walls you should install housewrap. The purpose of housewrap is to prevent moisture from coming in contact with the wall sheathing and entering the wall cavity, while allowing moisture vapor to escape. If properly sealed, housewrap can also act as an air barrier that improves the energy efficiency of your home by reducing drafts.

There are several different types and brands of housewrap on the market and they are not created equal. Some are made of a weaker woven and perforated plastic, while others use a stronger material with microscopic pores. These different characteristics

Housewrap Installed

impact how strong the material is and how effective it is at keeping out moisture. Based on my experience I strongly recommend using DuPont™ Tyvek®.

Housewrap is lightweight and comes in a long roll that makes it easier to install than tar paper. The fasteners used to attach the housewrap to the sheathing vary, based on the brand or type of housewrap. Generally, woven housewrap requires the use of plasticap or simplex nails, which are the same as those used to attach the tar paper to the roof. With stronger housewraps, staples can be used which can speed up the installation. Read the instructions for your particular brand of housewrap to verify that you are using the correct fastener and fastener spacing.

The housewrap should be installed on the lower portions of the house first, with the upper portions installed afterwards. This ensures that the upper portions overlap the lower portions, reducing the chance that water will get behind the housewrap. Once all the housewrap has been attached, you should seal the seams using a specialty tape made for housewrap.

Once the tar paper and housewrap are installed, your house is dried-in and should be mostly protected from the weather. If building outdoors, this can be a very satisfying milestone.

Proper installation of the windows and doors is important, as they are likely places for water to enter and damage your home

WINDOWS & DOORS

—

16

FLASHING

Flashing is an impervious material used to prevent water from coming in contact with wood or from entering a wall cavity. Flashing around windows and doors is generally a flexible, thicker material that is backed with adhesive to facilitate installation. Proper flashing is extremely important. If it is installed incorrectly, water will be able to penetrate the house and potentially cause extensive damage. Flashing can also be extremely difficult to fix because various layers of building material are added on top of it.

FLASHING AND INSTALLING WINDOWS

The first step to flashing and installing a window is to cut the housewrap for the window opening. This cut is in the shape of an upside down martini glass so that the housewrap can be wrapped around to the inside of the window opening on the sides and bottom. The top edge does not have a flap so that any water that might happen to run down the sheathing cannot get trapped in the housewrap.

Since any water that gets behind a window would come to rest on a horizontal surface, the window sill is the most vulnerable spot in a window opening. To reduce the chance that water will gather on the sill, cut a thin piece of lap siding the same size as the sill and attach it with the thinnest edge facing out. This will effectively slant the sill, reducing the chance that water will pool.

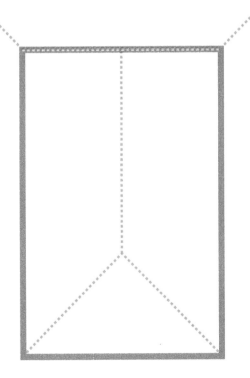

Cut Pattern for Housewrap over a Window

Housewrap Cut for Window

The housewrap flaps along the sides and bottom can then be wrapped around and attached on the inside of the house with staples.

Next, make two diagonal cuts in the housewrap starting in the two top corners. The cuts should be approximately 6 inches long, at 45 degree angles away from the window. This will form a flap that will overlap the top strip of flashing once installed. Temporarily tape this flap up and out of the way.

Since material higher up on the house should be overlapping lower materials, flashing is installed at the bottom first, with subsequent layers added directly above the previous piece.

Next, cut flexible flashing to the size of the window sill plus an additional 12 inches, so that it can be run up each side of the opening 6 inches. The flashing backing is then removed to reveal the adhesive and it is put into place so that the flashing edge is even with the inside edge of the window framing.

Cut slits in the flashing so that it can be wrapped around the exterior of the house. Note that some products do not require slits to be cut and the flashing can instead be stretched into position against the exterior.

Next, the window is inserted into the opening and a single screw is added to the top

Cut Siding for Window Sill

Flexible Flashing Laid Out

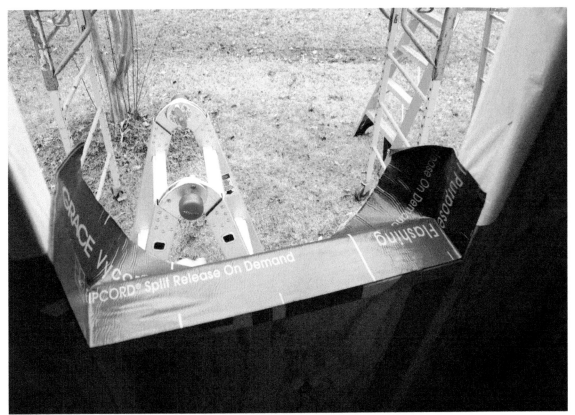

Flashing Partially Attached to Sill Plate

Flashing Attached to Sill Plate

right to hold it in place. Be sure that the flap made in the housewrap earlier is clear of the window flange. Adjust the window until it is square and plumb and attach the remaining screws to secure the window. Verify that the window operates properly. If it does not, the window is likely not square and will need to be corrected before proceeding.

Finally, flexible flashing is added to the outside of the window. First, it is applied to the sides of the window, extending beyond the top and the bottom by about 6 inches. Next, it is added to the top of the window under the housewrap flap, also extending beyond the edges by about 6 inches on either side. The flap edges should then be taped down over the flashing with housewrap tape.

No additional flashing is to be added to the bottom, since that could inhibit the escape of water if it does make its way down to the sill.

INSTALLING THE DOOR

The same technique that was used to create a sill for the windows is also used for the door. The only difference is that instead of using a tapered piece of wood on the sill, a ½ inch piece of plywood is used. On the windows, the function of the tapered piece was to angle the sill to encourage water runoff. With the door, however, an angled piece of wood is undesirable since the threshold of the door must be placed on a flat surface to evenly distribute any weight it might bear. The purpose of the plywood is to elevate the door a small amount to ensure that there is enough room under the door for the flooring and any rugs that might be placed on it. Some manufacturers build their sill so that it is thick enough and no extra plywood is required. Inspect your door's sill and compare it to the thickness of all the potential flooring material to determine if the extra thickness is required.

Once the sill is flashed, the door is stood up in the opening and attached as per the door's specific instructions. This procedure is usually very similar to how the windows were installed. First, a single screw is used to hold the door in place. Then, the door is squared up and tested to ensure the door operates as expected. Finally, the remaining screws are installed.

OTHER FLASHING

A piece of metal flashing is also nailed above the fender. This piece may need to be contoured if the fender is rounded. This can be accomplished by cutting several small slits at the spots where a bend is needed. Flexible silicone caulk should be applied liberally behind and underneath this flashing.

Flashing Attached to Sill Plate

Flashing Over Fender

The material and profile of your siding
greatly determines the final look and
feel of your house

EXTERIOR SIDING

There are several different options for exterior siding on a tiny house. A popular siding material for tiny houses is cedar which is lightweight, durable and attractive. Cedar is also available in different profiles. While lap siding has a more traditional look, shiplap siding can have a more modern appearance. Different profiles are available in different regions based on the local styles, so visit your lumber mill to determine what is available. Many current residential applications of lap siding use HardiePlank®, which is made with a concrete material. While great for a traditional home, HardiePlank® is not suitable for a tiny house due to its heavier weight and somewhat brittle nature.

The trim and siding should be painted or stained on both sides prior to installing. This will increase the durability of the material by protecting the back side in the event that water gets behind it. It will also prevent unstained or unpainted portions of the boards from being exposed when the wood expands or contracts. Paints and stains are also more consistent when applied to a horizontal surface, as the material is less inclined to drip or run.

Next, mark the walls with horizontal lines to line up the top edge of each piece of siding. These lines are made by measuring up from the bottom edge of the house at the corners and then using a 'chalk line' to connect the measured points. It is important that each line be measured and marked from the same starting point, such as the bottom edge of the house, and not from a previous line or from the top edge of a previously installed piece of siding. If the measurements are not taken from the same starting point, it is possible

Trim and Siding Completed

for slight imperfections in the measuring process or in the wood to compound resulting in significant differences by the time you get to the top of the wall. By measuring from the same point for all of the boards, a slight mistake in the placement of the chalk line for one board will not be carried over to the next.

Vertical lines should be marked on the exterior of the walls to indicate the location of each of the stud centers for nail placement. An easy way to make this mark is to take a measurement from the inside of the house to determine a stud's location. Then, adjust the measurement to account for the thickness of the sheathing material on the end of the house from which the measurement was made and use that measurement to mark on the exterior of the wall. Use a level to extend that mark up the side of the house. I find that a sharpie works best for making these lines on the housewrap, so that they are easily visible.

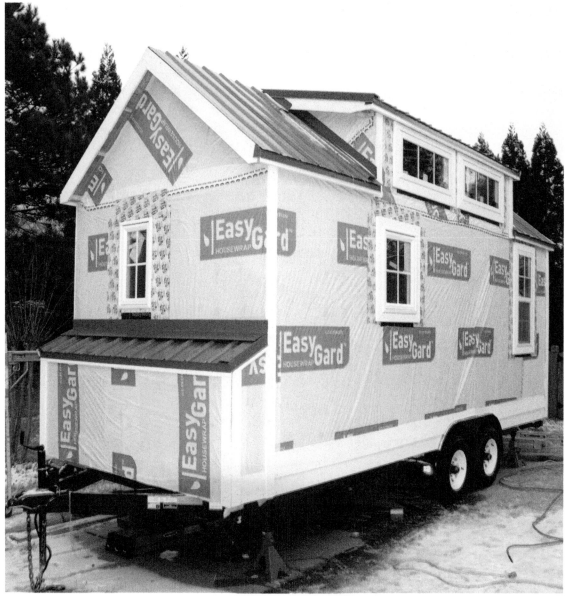

Trim Installed and Starting on the Siding

TRIM BOARDS

The next step is to install the trim boards around the doors, windows, and on the edges (corners) of the house. Trim boards are generally not larger than 1x4 inches; however, larger boards may be needed to avoid small, unworkable gaps between trim pieces. For instance, if a window is 8 inches from the edge of the house, using 1x4 trim, which is 3½ inches wide, along the edge of the house and around the window would result in a one inch gap between the two trim pieces. It is difficult to fill such a small gap with siding, as siding that is cut that narrow is hard to properly attach without it breaking or splitting. In a case like this, it may be better to use a wider piece of trim along the edge of the house to avoid the gap or to fill the gap with a cut piece of trim.

Any horizontal trim boards, like those above and below a window, should also have a piece of Z flashing installed with them. This prevents water from resting on the horizontal surface or from working its way behind the siding. For the trim on the top of a window, the Z flashing is installed above the trim and for the trim below a window, the Z flashing is installed below the trim.

The trim and siding is attached using 8D (2½ inch) spiral shank galvanized or stainless steel nails. Stainless steel is not as strong as regular steel, but for siding its strength is more than adequate. It will, however, have a much shinier appearance, so the nailheads will be more visible. Due to this visibility, it is a good idea to ensure all the fasteners line up. These are well suited if you want a more modern look, where visible fasteners may be desirable.

SIDING BOARDS

Once the trim is complete, the siding is then installed from the bottom up. The nails used to attach the siding should be driven into the studs, as indicated by the vertical lines you made previously. In some cases, holes may need to be pre-drilled in the siding to minimize cracking or splitting. This is usually done near the ends or on smaller pieces. The siding boards should fit snugly up against the trim pieces. Each edge of the siding will need to be cut to properly square it, as wood siding from the mill will rarely come with squared ends.

If the length of a wall is too long to use a single piece of siding, multiple pieces can be butted together. The end of any board that will butt up to another board should be cut at a 45 degree angle. This allows the boards to overlap, so that the sheathing and housewrap behind the joints are not visible if the siding were to contract.

If working alone, it may be difficult for you to hold a long piece of siding and nail it at the same time. To provide an extra 'set of hands,' a platform for the siding can be created from a few pieces of wood and a painter's extension pole. One end of the siding can rest on this platform while the other end is nailed.

FASCIA BOARDS

The next pieces of trim to be installed are the fascia boards. Fascia boards are 1x6 inch lumber that is attached to the edge of the roof overhang. When installing the fascia boards, be sure that no part of the board extends above the plane created by the top surface of the roof sheathing. An easy way to ensure the proper placement of these boards it to take a 2x4 and temporarily lay it on the roof sheathing so that it extends over the edge of the roof. The fascia board can then be placed so that it is snug up against the 2x4, ensuring that it is not installed too high and will not interfere with the metal roofing later.

Handmade Tool to Help Install Siding

Siding Installed on the Side of the House

Siding Installed on Gable End

While metal roofs are not as
commonly used on traditional houses,
they provide the best durability for
the high winds experienced by tiny
houses

ROOFING

—

On traditional homes, there are several different types of materials used for roofing, including asphalt shingles and metal panels. While asphalt shingles are the most common material, they are not very suitable for tiny houses because of their weight and durability. Metal roofing however, is easy to install, extremely durable, lightweight, and relatively inexpensive. For these reasons metal roofing is recommended.

The first component of a metal roof that is installed is the drip edge. Drip edge is a painted piece of metal that overlaps the bottom edges of the roof and covers the top portion of the fascia boards. It has a small bend at its base to allow water to fall away from the house. The drip edge is cut to size with steel snips and attached with roofing nails, using either a hammer or a pneumatic roofing nail gun. Drip edge should be wrapped around any edges of the fascia board by a few inches. This helps to keep water from getting behind it, as well as to more strongly secure it to the house.

Once the drip edge is in place, the roof panels can be installed. The roof panels often need to be cut to fit properly. Before cutting and securing any panels though, it is very important that you verify that the humps or raised portions of those panels will not interfere with the trim pieces that will be installed later. For example, gable trim flashing, depending on the profile, may need to make contact with the flat portion of a panel at 5 to 6 inches from the edge of the roof. If a hump or raised portion of the panel happens to be within this range, the panel and all attached panels will need to be shifted. This can be a very time consuming and frustrating mistake.

Drip Edge

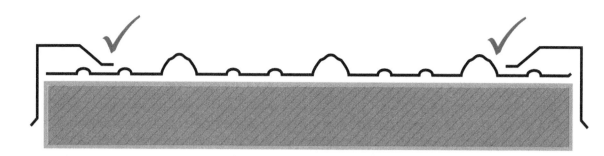

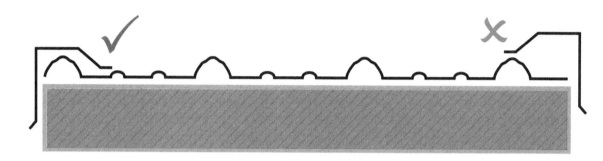

Roof Panel Properly Aligned, Roof Panel Improperly Aligned

Roof Panel Trimmed to Fit

CUTTING METAL ROOF PANELS

There are four different tools that we use to cut metal roof panels, each with their own advantages and disadvantages. Often using a combination of the following tools is required.

UTILITY KNIFE

A utility knife can be used to help cut panels lengthwise. Use the blade to score the panel, as many as eight to ten times, and then bend the panel back and forth until it snaps. This can produce a very clean cut. While scoring the panel, it can be helpful to clamp on a straight edge to ensure the blade is run over the same line each time. Be sure to clamp the straight edge to the side of the panel that will not be used in case it scratches the panel.

STEEL SNIPS

Steel snips work great for smaller cuts on thinner metal. They can also be used to make fine adjustments and curved cuts. However, they do not perform as well for longer cuts on thicker metal. This is because, unlike some of the other tools, snips do not remove any metal as they cut to make room for the tool itself. With thinner metal this usually is not a problem since the metal can be twisted or pushed out of the way to make room. On thicker metal, like that used for roof panels, it can be difficult to make room for the snips and the user's hand once the tool cuts more than several inches.

To get around this limitation a rough cut can be made a small distance from where the final cut is to be made. A second, more fine, exact cut can then be made on a second pass using the snips. Since a majority of the unwanted metal will be removed on the first pass, during the second pass the last bit of metal can easily be pulled out of the way while the cut is made. An effective tool to use for the first pass of this technique is an electric grinder.

ELECTRIC SHEARS

Electric shears are used to cut thicker metal for long lengths. Shears cut somewhat similarly to snips, with the exception that they remove about 3/16 inch of metal as they cut to make room for the tool. Electric shears can cut along flat surfaces very quickly and cleanly, however, it can be difficult to use them to cut over the humps found in panels. Thus, they are usually only used to modify the width of a panel and not the length.

ELECTRIC GRINDER

An electric grinder is the most destructive of the tools listed here. Instead of simply cutting the metal, this tool grinds away a section of it. This process throws off sparks and requires a firm grip to keep the tool in place. Caution and safety equipment should always be used, including eye and hand protection. The cut edge that a grinder leaves behind is generally pretty rough. Since this tool is primarily used to adjust the length of a panel, the cut edge should always be hidden by a piece of flashing or the ridge cap. With a grinder, multiple panels can be cut at the same time. When cutting a panel, be sure to have someone hold the panel in place or use clamps to secure it. Also be sure to remove any metal filings that may be clinging to it, as they will quickly rust and stain the panel.

Once the panels are cut, pre-drill the holes for the screws. Consult your roofing manufacturer for the suggested screw placement and spacing. I put screws on both sides of each hump on the panel along the bottom edge. I then put screws on both sides of alternating humps, spaced approximately 18 inches apart, beginning with the hump at the seam. Also check with the panel manufacturer to determine the minimum amount of overhang of the panels to determine how high up from the edge the first row of screws should be placed. The recommended amount of overhang for the panels we use is 1 inch, so I drill the first line of holes 2 inches up from the bottom edge. If a standing seam roof is used, special brackets are used instead of drilling through the panel itself. See your manufacturer for instructions.

Finally, butyl tape is placed on the inside of any humps that will overlap with another panel to make a seam. This step is often not required by the manufacturer unless the pitch of the roof falls within a certain shallow range, since water stays on a gently sloped roof longer. Roofs on tiny houses are so small that I recommend using butyl tape regardless of the pitch, since it will not take much additional effort or expense.

Once a panel has been prepared (cut, drilled and taped), it can be put in place and attached. Care should be taken since a panel can act as a sail on a windy day and pull or push someone off of a roof or ladder. Ensure that no one is standing below or around the installer while a panel is being mounted, since a panel can easily slide off of the roof before it is screwed.

When putting the panel in place, try to avoid sliding it, as this can easily scratch other panels or the drip edge. Also, ensure that the panel is square with the roof and that the amount of overhang across the entire panel is consistent. Since all the panels will be connected, a small difference on the first panel may end up being a considerable difference by the time the last panel is installed.

The roof panels are attached with color matched screws that have a butyl washer backing to eliminate leaking around the screw hole. As the screws are painted to match the panels, they should be obtained from the roofing manufacturer. Be careful not to over tighten the screws because doing so will weaken the seal and dent the panel. A good method to prevent over tightening is to adjust the drill torque setting so that the drill will stop driving the screw once it gets to the proper tightness.

Installed Roof Panel

Roof Panel Being Installed

Sidewall Flashing Being Installed

Ladder with Padding

Finally the roof trim pieces, including the gable trim, sidewall flashing, end wall flashing and ridge cap are installed. These pieces can generally be cut using steel snips. Occasionally, to cover longer lengths, individual pieces of trim may need to be overlapped. In these cases, the pieces should overlap by at least 6 inches. Be sure the pieces closer to the front of the trailer overlap the pieces farther back so that the seams will not catch the air as the house is driven down the road. As with the other components, check with the roofing manufacturers for installation details including fastener spacing.

Note that depending on the complexity of your roof, some pieces may need to be installed out of order. For instance, if you have a dormer, it may be easier to work on the trim under the dormer overhang before installing the last rafter on that overhang. Without that rafter being installed, you will not be able to install the panels above the dormer. So in that case, some of the roofing trim work will be completed before all the panels are in place.

If scaffolding is not available while installing the roof, add padding or guards to any ladders to prevent scratching the metal panels.

The pictures below illustrate the steps to finish the ridge cap ends.

End Marked

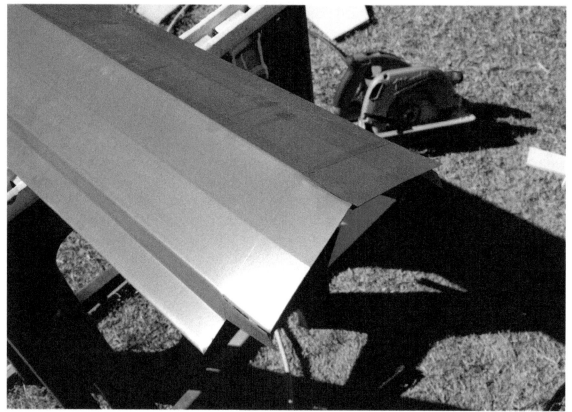

End Cut

End Cut and Folded Down

Sharp Edges Trimmed

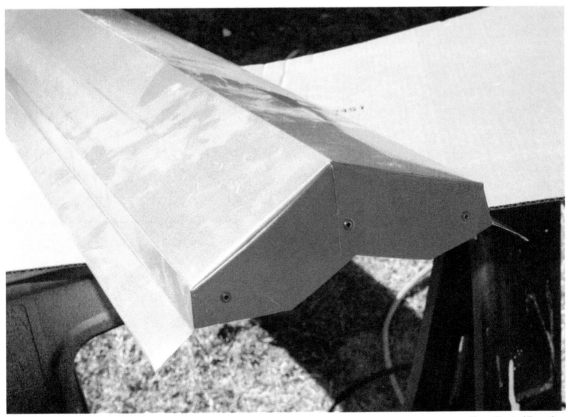

Flaps Riveted Together

Many people have concerns about
plumbing a tiny house, but it is
actually one of the easier tasks

PLUMBING

—

PLUMBING COMPONENTS

The first step to learning about the plumbing system of a house is to understand the individual components. A plumbing system is composed of two different sections, the supply lines and the waste lines.

SUPPLY LINES

The supply lines, sometimes called distribution lines, bring fresh water into a house. For supply line plumbing material there are several different options including copper, PEX, and CPVC.

Plumbing Parts

COPPER

Copper has been used for plumbing the longest of all the materials listed. However, there are some distinct disadvantages to using copper in your tiny house.

First, copper is the most expensive of the materials listed. Since a tiny house will require so little plumbing material, this may not be a big concern with your budget. However, its higher value and ease of selling makes it very prone to theft. Thieves have been known to do thousands of dollars in damage just to get to a very small amount of copper. Also, because of its higher value some foreign manufacturers have been known to mix in other metals with the copper. This can result in the copper leaking, as the other metals corrode over time.

Copper is also harder to work with and install. It is very rigid which can make it difficult to route through walls. Any holes cut in the studs will need to line up precisely for the piping to fit through. Finally, copper also requires the most skill to assemble and it can take a bit of practice to become proficient at it.

Because of these negatives, we do not recommend using copper as your supply line piping material.

PEX (CROSS-LINKED POLYETHYLENE)

PEX is a newer plumbing material known for being highly flexible. It is less expensive than copper, but more expensive than CPVC.

In a conventional home, the flexibility of PEX can make for a much easier installation and result in fewer fittings, which are the pieces that hold the piping together. In a tiny house, however, the plumbing is very limited and confined. That amount of flexibility is not as useful or needed.

PEX is very easy to install and work with, although it requires a specialized tool that costs approximately $70 to connect the piping to the fittings.

While PEX has several advantages over copper, its biggest advantage of being highly flexible is lost on tiny houses.

CPVC (CHLORINATED POLY (VINYL) CHLORIDE)

CPVC is both inexpensive and easy to work with.

While not as flexible as PEX, CPVC is flexible enough to bend slightly within the walls making installation easy. The pipes are also easy to cut and join using a specialized liquid cement. CPVC will not corrode and is suitable for higher temperature water.
For these reasons, CPVC is the recommended plumbing material.

Copper Piping

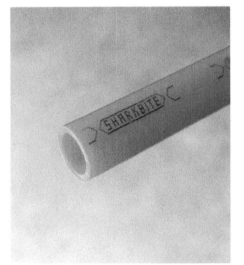

PEX Piping

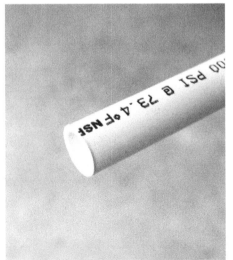

CPVC Piping

There are different types of water heaters, including tank and tankless.

Tank water heaters are the predominant style of water heater in the USA. They work by storing water in a tank and using a heater element to bring the temperature of the water up to a set point. When hot water is needed, it is pulled from the tank of pre-heated water. There are several downsides to this approach. First, the tank can be very heavy since it is full of water. In a tiny house, a smaller water heater can be used, but you run out of hot water much quicker. Second, the heater element is always running and keeping the water at the set temperature regardless if the water is being used. If you went out of town for a week and forgot to turn your water heater off, it would be keeping your tank hot unnecessarily.

Tankless water heaters on the other hand do not have a tank and only heat the water as it is needed. These are sometimes called on-demand water heaters. This approach is much more efficient and results in a water heater that is lighter to transport. Tankless water heaters work by using high-powered burners to quickly heat up water as it flows through the heater.

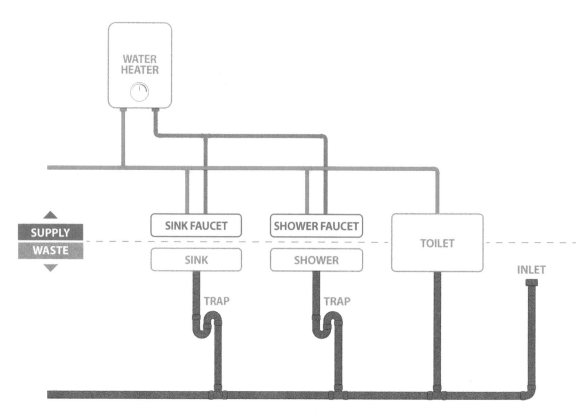

Plumbing Overview

For the waste or drain lines, PVC piping is used. PVC piping is similar to CPVC piping except that PVC is not suitable for hot water. You will typically see PVC piping used for drain lines and service lines (supply lines before the water is heated up with the water heater).

Since the PVC drain lines will be located below the trailer and thus subject to road debris, the thicker variant of PVC piping called 'Schedule 80' is recommended.

INSTALLING THE PLUMBING SYSTEM

The plumbing system in a house is installed in two phases. The first phase is called the rough-in, and the second phase happens when the plumbing installation is completed. In this overview, I will describe the installation of a CPVC and PVC based plumbing system.

SUPPLY ROUGH-IN

The plumbing supply rough-in involves installing all the pipes in the walls of the house, so it must be done before the insulation and interior siding are added.

The piping should be laid out and routed such that there is a hot and cold line at each faucet and shower location, and a single cold line at the toilet location.

For the piping you will use ½ inch pipes, which for CPVC is measured from the outside diameter of the pipe. The piping is cut to the desired length using a specialized tool called a PVC or CPVC pipe cutter. A hacksaw or miter saw can also be used, but these tools frequently leave behind burs that will need to be removed before the pipe is put into a fitting. A utility knife or a deburring tool can be used to accomplish this.

Next, apply a primer to both the exterior end of the pipe and the interior of the fitting that you plan to attach to the pipe. The primer will help soften the pipe to help seat it within the fitting. Next apply CPVC cement on top of the primer on both the pipe and the fitting. The primer and cement are applied using a dauber that is built into the cap of both products. Be sure that the cement selected is specific to CPVC, as it is not interchangeable with PVC cement.

Firmly press the pipe into the fitting. Once the pipe is pressed all the way in, turn the pipe a quarter of a turn to help spread the cement evenly across the entire connection. The primer and cement will react with each other and melt the pipe and fitting together for a permanent bond. Hold the pipe and fitting together for 30 seconds to ensure that the heat from this reaction does not push the connection apart. Repeat this for all of the connections.

For brass threaded connections, like those found on the shower assembly, a threaded CPVC adapter is used. First, the threads of the adapter are covered with a thread seal or plumber's tape. The adapter is then screwed into the shower assembly. This adapter provides a small stub of CPVC that a pipe can then be cemented directly to as described above.

At the locations where the pipes will eventually be connected to the plumbing fixtures, the pipes should be left with ends extending out of the wall cavity a few inches. Unglued caps should be placed over the open pipe ends to stop insects from making the pipes their home during construction.

Optionally, if you are not confident in your plumbing skills, all of the pipes and fittings can be cut and assembled without cement so that you can minimize the impact of any mistakes. Then, only after the system is completed would you need to go back and make it permanent by using the primer and cement. Once complete, double check that every fitting joint has been cemented.

WASTE ROUGH-IN

For the waste lines, 2 and 3 inch PVC piping is used. For PVC, this measurement is the inside diameter of the pipe. PVC pipe is connected together much like CPVC except a different PVC specific cement must be used. Due to the size of the PVC pipes, they will

Supply Entry and Shower Hookup

Hot and Cold Pipes for Kitchen Sink

Supply Entry into House

Cold Pipe for Toilet

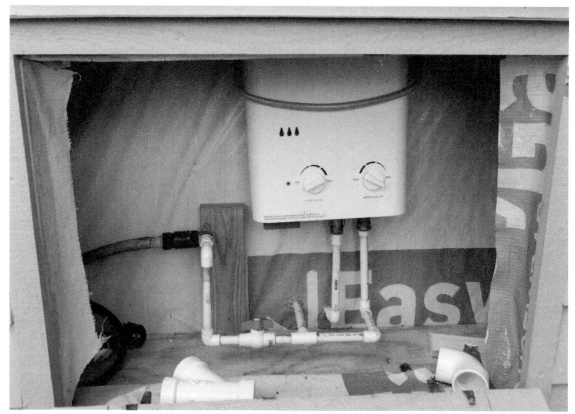

Hot Water Heater

Drainage Pipes under Trailer

need to be cut with either a hacksaw or a miter saw. You will need to remove any burs resulting from the cutting process.

Waste pipes need to slope down away from their source. For every 1 foot of pipe, it will need to drop between ¼ inch and ½ inch. This is to make sure that waste does not stay stagnant in the pipes.

Each drain source also needs to have a trap. A trap is a bend in a pipe where water gets trapped and prevents sewage gas from entering back into the house. A kitchen sink has a trap directly below it that is visible in the cabinet underneath. A toilet has a trap built into the porcelain base of the toilet itself. A shower, however, does not have room to have a trap inside the house, since the base of the shower is resting on the floor. For this reason you must add a trap into the waste plumbing under the house.

An air inlet line must also be included in the waste system. When waste travels out of the system, it can create a vacuum. An air inlet prevents this vacuum effect. The inlet is usually vented to the outdoors as high as possible, since the inlet cannot have a trap and sewer gasses may escape from it. In a conventional house, the air inlet line usually extends through the attic and exits the roof. It is easily identifiable as it does not have a cover or valve to prevent rainwater from entering it. In a tiny house, it may be difficult to vent the drainage lines through the roof. Instead, this line can be run up within a wall and vented out the side of the house. It should not be vented to the same side as the front door.

Drainage Pipes under Trailer

Finishing the plumbing is one of the very last steps in building a tiny house. This is because it cannot be done until all the cabinets and countertops are installed.

To finish the plumbing, you will trim back the excess piping that was previously left protruding from the walls and using the same methods discussed earlier for attaching fittings, attach valves to the end of the pipes. These valves will attach to the CPVC on one end and have a 3/8 inch compression outlet on the other end. A flexible water connection hose is then used to connect the valves to the faucets.

OFF-GRID PLUMBING

COLLECTION

Rain water collection is common in off-grid systems, but may not be as suitable for a tiny house unless supplemented with other sources.

The catchment area of a tiny house alone is probably not large enough to provide enough water for its occupants year round. Unless you live in an area with a high annual rainfall or use very little water, your demand will outstrip supply. To determine how much water you can harvest, use the following formula:

Catchment Area (square feet)	x	Rainfall Depth (inches)	x	0.623 Conversion Factor	=	Harvested Water (gallons)

This formula should be calculated for each month, since average rainfall will vary from month to month.

When calculating the amount of water you will need, be sure to consider all or your activities that use water. Below are a few of the common activities and their typical consumption.

Activity	Typical Consumption
Drinking	1 gallon per day
Shower	2 gallons per minute
Toilet Flush (if using a standard flushable toilet)	1.6 gallons
Washing your hands	1 gallon per minute

Assuming your usage is 15 gallons per day and the dimensions of your tiny house are 8 feet by 24 feet, then you would require approximately 4 inches of rainfall each month. This exceeds the average rainfall in the USA for 9 out of every 12 months.

While rainwater catchment in a tiny house is possible, you will want to have a backup or supplemental water source.

COMPOSTING TOILETS

One of the most asked about topics of off-grid living is how to properly dispose of human waste. One popular option is to use a composting toilet. There are several different commercial and do-it-yourself (DIY) options available for composting toilets. While the procedures for each might vary slightly, the overall concept is the same.

Many of the most popular composting toilets in use today are called dry toilets or Urine-Diverted Dry Toilets (UDDT). This is because when they are used, the liquid waste is separated from the solid waste with a diversion mechanism.

The liquid waste or urine is disposed of frequently. In healthy humans urine is almost always sterile so there is not a high health concern around handling it. Urine can be high in both salt and nitrogen, so if it is to be used in gardening, it should first be diluted with water at a ratio of 1:10. Otherwise, is can be disposed of at the base of a tree.

Solid waste is treated within the composting toilet with a compost material like peat moss, coconut fiber, sawdust, or even some types of cat litter. It is then left to compost in place with a vent on the compost chamber, removing any odors and reducing humidity. As the waste is composted, its appearance and smell changes to that of dirt. Because the size of the waste is also reduced, the toilet may only need to be emptied once every couple of months. For instance the manufacturer of a popular dry composting toilet recommends the solid portion of the toilet be emptied after 60 to 80 uses.

An alternative to using a composting toilet is to use a black water tank where all the waste, both solid and liquid, goes into the same chamber. This concept is more comfortable to many people because it closely resembles a typical flushing toilet. This is the approach that is used in many RV's, since an RV can be pulled up to a dump station where the tanks can be emptied. However, since most tiny houses will not be moved frequently, the tanks would need to be portable. If you are contemplating this option I highly suggest you reconsider composting toilets. The logistics of frequently moving and emptying heavy waste tanks is far more difficult and unappealing than occasionally handling compost.

Electrical is a large, detailed topic. Fortunately, only a very small subset of knowledge on the subject is required to wire a tiny house

ELECTRICAL
—

ELECTRICAL COMPONENTS

Just like with the plumbing system, the first step to learning about the electrical system of a house is to understand the individual components.

WIRE

Residential wiring, sometimes called by the brand name Romex®, is sheathed to protect it in the walls of the house. It comes in different gauges, or thicknesses, as well as having a different number of wires within the sheathing. The product packaging will indicate these two items, separated by a slash. For instance, wire that is 12 gauge and has two wires plus the ground wire within the sheathing is called '12/2' wire. The type and number of electrical devices connected to a circuit determines the gauge of the wire that should be used. Please note, the smaller the gauge number, the thicker the wire. In a typical tiny house, 12 gauge wire can be used throughout. Standard outlets and switches require only 2 wires, plus the ground, in the sheathing. 3-way switches require 3 wires.

The diagram bellow illustrates a '12/2' wire, where there are 2 individually sheathed wires in addition to the ground wire, which is just bare copper. The white wire is the neutral wire and the black wire is the hot wire. The ground wire and the neutral wire are closely related, as they are both connected to the earth at some point. For safety reasons they are kept separate throughout the entire system until they are joined together in the main breaker box or panel. The hot wire is the wire that has electric potential relative to the ground and neutral wires, so it is the one that can shock you. In unique circumstances, the ground and neutral wires can also shock you, so always use caution. Fortunately, when working on the electrical system of a tiny house it is simple to disconnect it from its power source to avoid getting shocked.

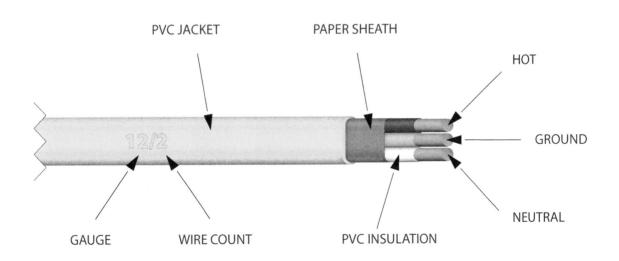

Sheathed Wire

OUTLET BOXES

Outlet boxes are boxes that will hold the switches and outlets in a wall. They come in different sizes, depending on the number of outlets and switches they will contain, called 'gang'. For example, 1 gang, 2 gang, etc.

OUTLETS

Outlets provide a point for which electric current can be drawn from the electrical system to power household items.

SWITCHES

Switches are used to interrupt or divert the flow of electricity. Only a single wire needs to be interrupted to break a circuit, so only the 'hot' wire is hooked up to a switch.

BREAKERS

A breaker is an automatically operated switch that turns itself off if it detects an overloaded circuit. A breaker is sized by the amount of electric current it will allow to pass through before turning itself off, such as 15 amp or 20 amp. When electric current passes through a wire, the wire will heat up. If too much current passes through a wire, it can heat up too much and melt and short, possibly resulting in a fire. A breaker is used to prevent this by only allowing an appropriate amount of current to pass through the wire that is attached to it.

Wire Gauge	Breaker Amps
14	15
12	20
10	30

If the house is wired with 12 gauge wire throughout, then all the breakers would be 20 amp.

A breaker services a section of a home. For instance, in a traditional home there might be one breaker for the master bedroom, another one for the family room and so on. In a tiny house you might have one breaker for the right side of the house, another for the left side, and additional breakers for appliances that have a larger electrical draw like the air conditioner or refrigerator.

BREAKER BOX

A breaker box, also called a panel, is a steel box that holds all the breakers. Electricity enters a house at the breaker box and it serves as a distribution point for all the power in your house.

Now that you know about the individual components, it is important to know how they all fit together to make up the electrical system. First, as previously mentioned, the main electrical supply enters the house at the breaker box. If you were to take the cover off of a breaker box in a traditional home, it can be overwhelming trying to understand what is going on in there. What is happening is actually quite simple, it is just happening over and over which makes it look more complicated than it is.

Breaker Box in a Conventional Home

Inside the breaker box, the breakers, just like switches, are only connected to the 'hot' wires. Since there is only a single 'hot' wire from the power supply, it is attached to a 'bus' (or metal strip) that all the breakers are connected to. This allows the single hot supply to be easily shared with many breakers. The supply 'ground' and 'neutral' wires are also connected to different metal strips to facilitate splitting those wires among all the circuits. There are no breakers for the 'ground' and 'neutral' wires and the breaker box acts only as a pass-through for those wires.

Regardless of the number of breakers, there is only ever a single supply which is 3 individual wires (4 if there is 240V service). Then there will be 3 additional wires that correspond to each breaker. So in a conventional home with 20 breakers, the breaker box will have 63 wires coming into it. While what is happening inside the breaker box is rather simple, you can see how it can start to look really complicated.

Each sheathed wire that leaves the breaker box services a section of the house. Since each section of the house will probably have more than one outlet or light, the electricity must be shared across multiple endpoints (outlets, switches, and lights). To facilitate this, the outlets have two connection points for both the hot and the neutral wires. One for the power supplying the outlet, and one for the power leaving to supply the next outlet or switch. These two connection points are bridged together so they are equivalent. There is no need to differentiate which one the supply is connected to and which one the outgoing power is connected to. You will, however, need to make a distinction between which side to connect the 'hot' or black wire to and which to connect the 'neutral' or white wire to. Usually there is a label on the back of the outlet identifying the hot side, and the silver screws are for the neutral connection.

When installing a switch, since it only needs to interrupt the hot wire, the neutral and ground wires simply pass-through the box. For that reason, a switch only has two connection points, one for the hot wire coming in and one for the hot wire going out. The orientation of the switch is important, since a switch in the up direction is associated with being on. For this reason the top of the switch is labeled.

3-way switches are used to control a single light from multiple locations. This can be useful if you wanted to control the main lights in your house from both downstairs and from the loft. For instance, if you wanted to turn the lights on when you got home and then off after you were in bed. A 3-way switch works differently than a standard switch. Instead of interrupting the power passing through it, it re-routes the power through a different path. To support this additional path, an additional wire in the sheathing is required (i.e. '12/3' wire). Diagrams found later in this chapter illustrate how 3-way switches in conjunction with 3-wire can be installed to provide this functionality.

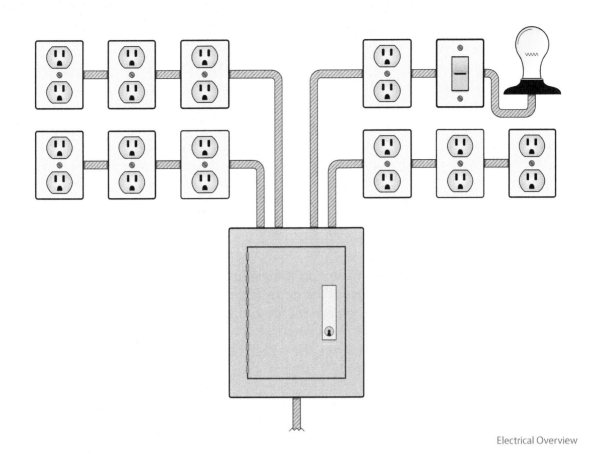

Electrical Overview

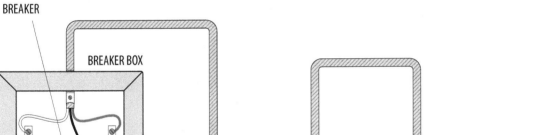

BREAKER

BREAKER BOX

POWER IN

NEUTRAL   HOT   GROUND

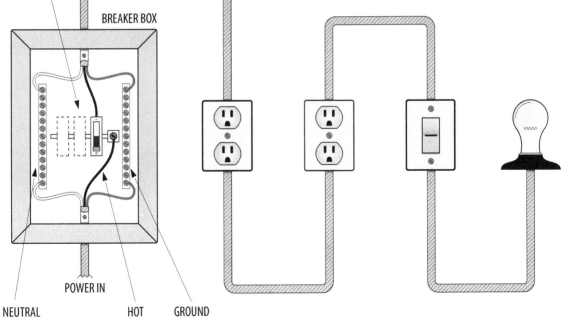

Breaker Box

INSTALLING THE ELECTRICAL SYSTEM

The electrical system in a house is installed in two phases. The first phase is called the rough-in and the second phase happens when the electrical installation is completed.

ROUGH-IN

The electrical rough-in involves installing the outlet boxes and the wires in the walls of the house; therefore, must be done before the insulation and interior siding are added. The first step is to install the switch and outlet boxes at the desired locations. If you have purchased plans for your tiny house, the plan package may include an electrical diagram that specifies the placement of these boxes.

The boxes come with small raised 'tabs' on their sides that allow for them to easily be offset, so they extend a ½ inch beyond the edge of the studs. If a thinner interior siding material is to be used, these raised sections should be removed using a utility knife. When installed, the boxes should not extend beyond the stud any further than the interior siding material. Otherwise, when the switch plate covers are installed, they may hover above the interior siding.

Switch boxes should be placed 48 inches from the subfloor to the top of the box. Outlets should be placed 16 inches from the subfloor to the top of the box.

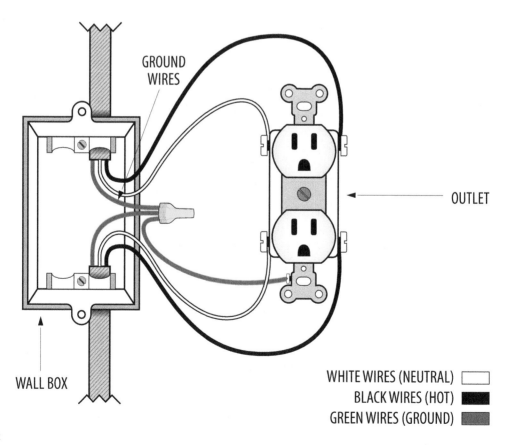

GROUND WIRES

OUTLET

WALL BOX

WHITE WIRES (NEUTRAL)
BLACK WIRES (HOT)
GREEN WIRES (GROUND)

Outlet

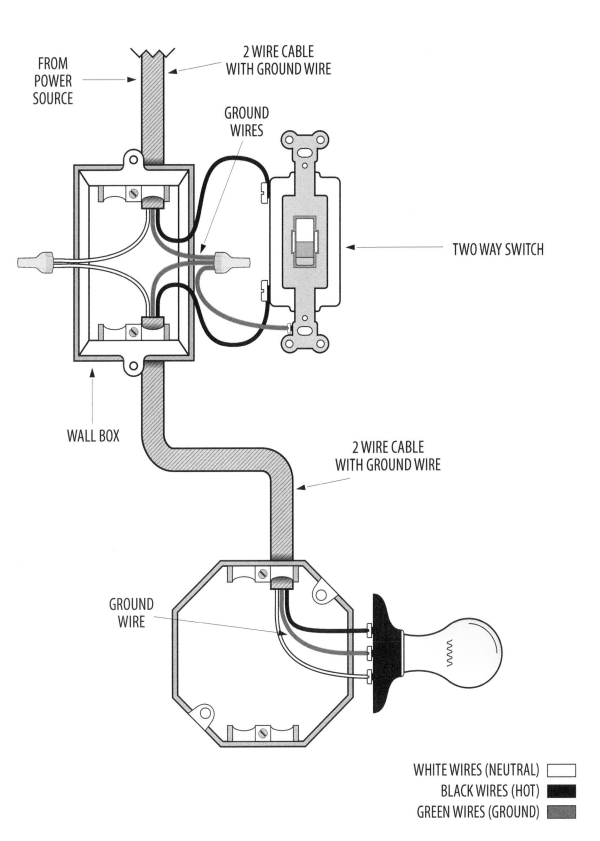

FROM POWER SOURCE

2 WIRE CABLE WITH GROUND WIRE

GROUND WIRES

TWO WAY SWITCH

WALL BOX

2 WIRE CABLE WITH GROUND WIRE

GROUND WIRE

WHITE WIRES (NEUTRAL)
BLACK WIRES (HOT)
GREEN WIRES (GROUND)

Standard Switch - Power Through Switch

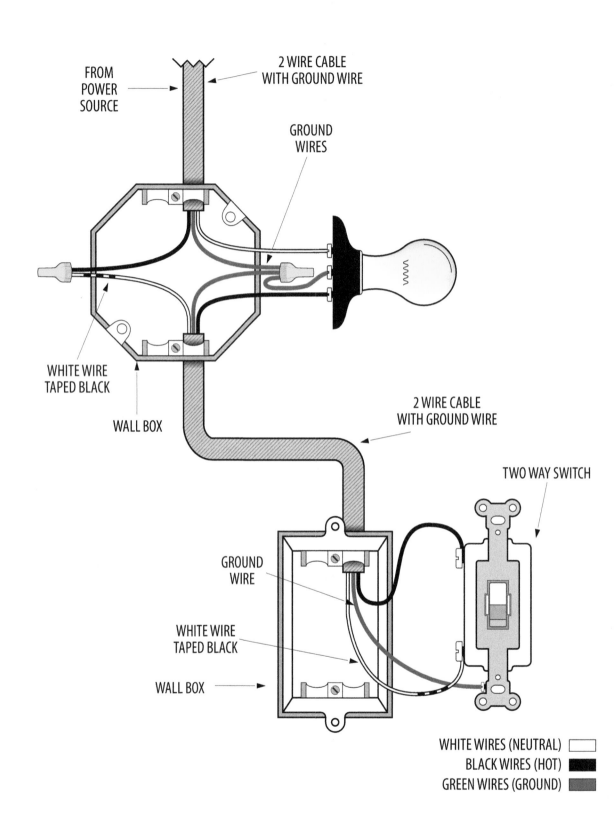

FROM POWER SOURCE

2 WIRE CABLE WITH GROUND WIRE

GROUND WIRES

WHITE WIRE TAPED BLACK

WALL BOX

2 WIRE CABLE WITH GROUND WIRE

TWO WAY SWITCH

GROUND WIRE

WHITE WIRE TAPED BLACK

WALL BOX

WHITE WIRES (NEUTRAL)
BLACK WIRES (HOT)
GREEN WIRES (GROUND)

Standard Switch - Power Through Light

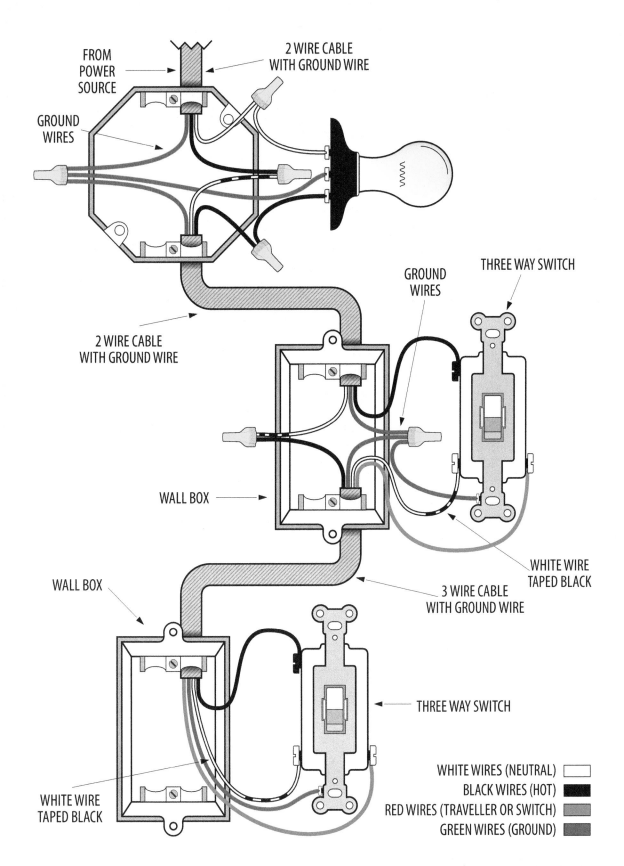

FROM POWER SOURCE

2 WIRE CABLE WITH GROUND WIRE

GROUND WIRES

GROUND WIRES

THREE WAY SWITCH

2 WIRE CABLE WITH GROUND WIRE

WALL BOX

WHITE WIRE TAPED BLACK

3 WIRE CABLE WITH GROUND WIRE

WALL BOX

THREE WAY SWITCH

WHITE WIRE TAPED BLACK

WHITE WIRES (NEUTRAL)
BLACK WIRES (HOT)
RED WIRES (TRAVELLER OR SWITCH)
GREEN WIRES (GROUND)

3-Way Switch - Power Through Light

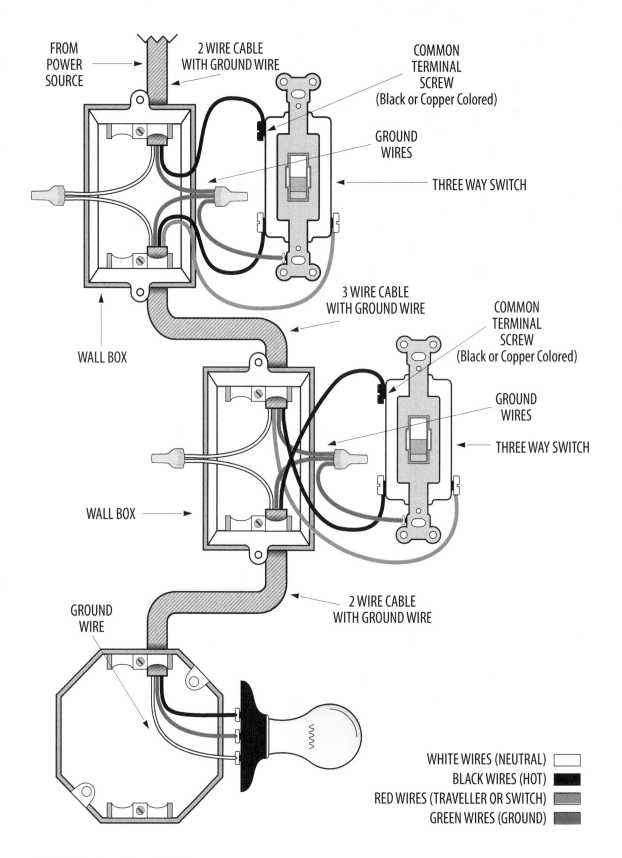

FROM POWER SOURCE

2 WIRE CABLE WITH GROUND WIRE

COMMON TERMINAL SCREW (Black or Copper Colored)

GROUND WIRES

THREE WAY SWITCH

3 WIRE CABLE WITH GROUND WIRE

COMMON TERMINAL SCREW (Black or Copper Colored)

GROUND WIRES

THREE WAY SWITCH

WALL BOX

WALL BOX

2 WIRE CABLE WITH GROUND WIRE

GROUND WIRE

WHITE WIRES (NEUTRAL)
BLACK WIRES (HOT)
RED WIRES (TRAVELLER OR SWITCH)
GREEN WIRES (GROUND)

3-Way Switch - Power Through Switch

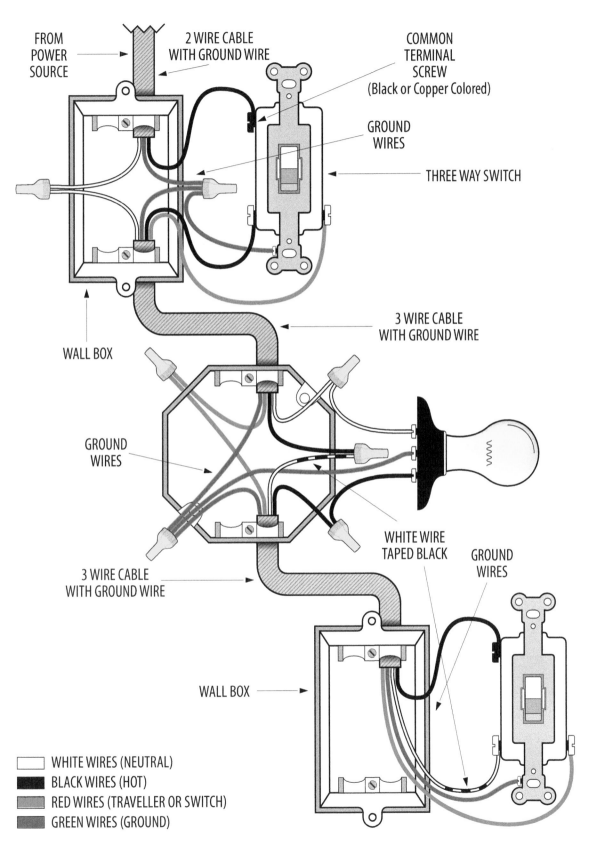

FROM POWER SOURCE

2 WIRE CABLE WITH GROUND WIRE

COMMON TERMINAL SCREW
(Black or Copper Colored)

GROUND WIRES

THREE WAY SWITCH

WALL BOX

3 WIRE CABLE WITH GROUND WIRE

GROUND WIRES

WHITE WIRE TAPED BLACK

GROUND WIRES

3 WIRE CABLE WITH GROUND WIRE

WALL BOX

WHITE WIRES (NEUTRAL)
BLACK WIRES (HOT)
RED WIRES (TRAVELLER OR SWITCH)
GREEN WIRES (GROUND)

3-Way Switch - Light Between Switches

Next, map the route of the wires starting at the location where the breaker box will be installed. Be sure that the wire's path will not interfere with the plumbing and that the electrical wiring is run separately from other wires such as cable and telephone lines. If narrow slat siding is being used in the interior, it is recommended that the wires be run at an angle through the walls instead of horizontally. Since metal protective plates will be installed on the studs at the locations of the wires, if the wires were to be run horizontally, these metal plates would line up to prevent a single piece of slat siding from being nailed to the wall. This can result in that row of the siding buckling or bowing.

Drill a ¾ inch hole in the studs to run the wires through. A metal nail plate should be attached to the outside of the studs, directly over where the holes are drilled, to prevent accidentally driving a nail into the wire later when installing the interior siding or hanging pictures.

The wires are then run from box to box. Approximately 8 inches of wire should be left for connecting to the switches and outlets later. This excess wire should be rolled up and tucked inside the box so that it does not get in the way. There can be no breaks in the wire other than those within the boxes. It is a fire hazard and against electrical code to have breaks in or connections between wires within the wall cavity.

Every effort should be made to limit the number of wires entering a single box. For a 1 gang box, try to have no more than three wires per box. All of the wires and the outlet

Dedicated Line for the Air Conditioner

or switch will need to fit inside the box later on. The more wires in the box, the harder it will be to make it all fit.

Once the interior siding has been installed, the electrical can be completed by installing all the switches, outlets, and switch plate covers. Review the diagrams in this chapter for details on how each of those are connected. To aid in fitting the wires and the component into the box, fold the wires together in an accordion shape, the same height as the interior of the box. This will not only make it easier to fit all the parts in the box, but will also reduce the chance of a wire pulling free and causing a failure due to excess tension.

Since the electrical wiring will not be tested until the wall sheathing is installed and the switches and outlets are connected, fixing fundamental wiring mistakes will likely require removing the wall siding. This can be time consuming and expensive. To avoid any mistakes, you may wish to install all the switches and outlets as part of the rough-in so thorough testing of all the wiring is possible earlier. The switches and outlets would then need to be removed, only to be installed again later at the appropriate time. While this may add on an additional day of labor, it can be well worth it if a mistake is discovered prior to closing the walls.

Proper electrical installation and wiring is extremely important, so I encourage you to

Switches and Outlets

do additional research or consult with a licensed professional if you are uncertain about any details.

OFF-GRID POWER

The electrical system we have discussed up to this point is an alternating current (AC) power system. AC electric is provided by utility companies and is the system found in almost every house. Direct current (DC) power, on the other hand, is a different type of power that can be associated with the electricity stored in batteries. There is no need to understand the details of the differences between the two power types for this overview, other than to understand that there is a difference and that they are not interchangeable. Devices are made to run with electricity of a specific power type. For instance, your toaster oven runs on AC power because the manufacturer knows that your house is powered by AC. Your phone is powered by a battery, so it runs on DC power. Since your house is powered by AC, the power needs to be converted to DC in order to charge your phone. This conversion happens with an AC to DC converter that is located in your phone's charger. Just like AC can be converted to DC, DC can also be converted to AC through a device called an inverter.

ENERGY COLLECTION

The primary component in an alternative energy system is the energy collector. The two most popular types of energy collectors are solar and wind, however, there are also other types including geothermal and hydro. The purpose of an energy collector is to collect and convert energy into a form that is easily used, usually DC power.

STORAGE

In a traditional home, power is being created and delivered by the utility company as it is needed. When power is collected from alternative sources, it is usually stored locally before being used. This is because alternative sources are not normally as consistent or predictable. For example, if the alternative source of power is solar, it may be cloudy one day or you may want to use power at night when solar energy is not being collected. By storing the collected energy, you are able to have a much more predictable power source. The energy collected is stored in batteries, so it is stored as DC power.

CONSUMPTION

Once the energy is stored, it can be consumed immediately or used later. The energy is stored as DC power, so if your house is wired for AC it will first need to be converted by using an inverter. The downside to using an inverter is that they are typically only 80% efficient, meaning 20% of the power is lost in the conversion process, typically as heat loss. If possible, it is more efficient to use DC directly. For instance, it would be better to use a DC adapter for your laptop and phone since they run on DC power. Otherwise, you would be converting your DC power from the battery bank to AC with the inverter only

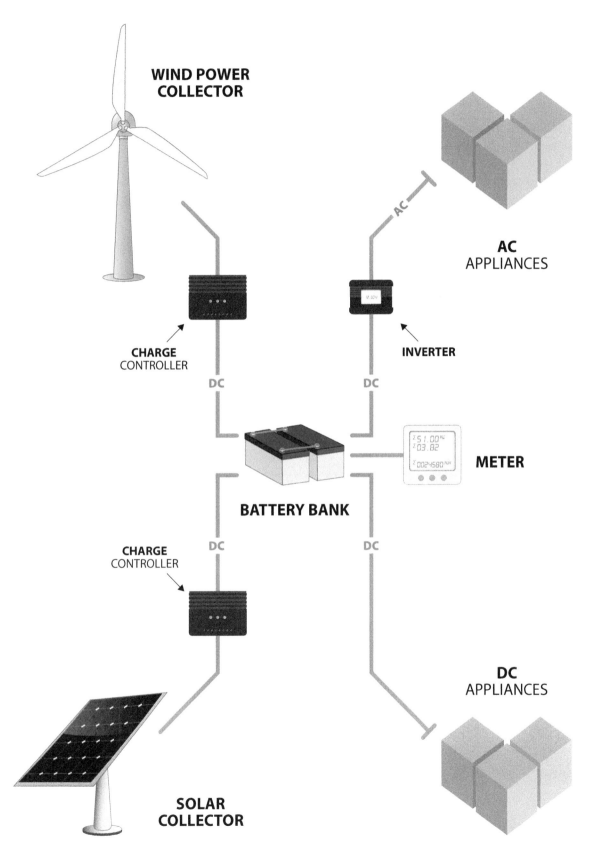

WIND POWER COLLECTOR

CHARGE CONTROLLER

DC

AC APPLIANCES

AC

INVERTER

DC

BATTERY BANK

METER

CHARGE CONTROLLER

DC

DC

DC APPLIANCES

SOLAR COLLECTOR

Alternative Energy System Overview

to have the phone's charger convert it back to DC.

There are also appliances designed to run on DC power. These are common in the RV industry since RV's are designed to be taken off-grid for a few days at a time and typically run on batteries while doing so.

DC power requires a separate set of wires than those used for the AC power in a house. If you plan to use an alternative power source, it can be beneficial to wire the house for both AC and DC. If the house is already wired for only AC, you will just need to size the energy collection and storage systems larger to account for the extra waste from converting all the power to AC.

SYSTEM SIZE

Many factors go into determining the size of a collection and storage alternative energy system. The two biggest factors are how much energy you use and how much energy you can collect.

If you are already living in a tiny house and plan to convert to an alternative energy system, determining how much energy you use is easy. That information is provided on your electric bill. If you are not already living in your tiny house, then you do not want to use your current electric bill as a benchmark, since your energy consumption will likely be reduced dramatically after moving. In this case, you will need to complete an energy audit and estimate your future usage.

Appliance	Running Wattage (Watts)
Light Bulb - 60 Watt	60
Light Bulb - 75 Watt	75
Clock	10
Refrigerator/Freezer	700
Toaster	850
Microwave Oven - 625 Watt	625
Coffee Maker	800
TV - 32 inch LED	55
Iron	1200
Hair Dryer - 1250 Watt	1250
Laptop Computer	30-70
Mobile Phone	5-12
Video Game Console	100

Typical Energy Consumption

An energy audit consist of identifying each powered item you use during the day, how long you use it, how much energy it consumes and then add all those items up. This will give you a realistic idea of your total energy usage. An example would be if you use a coffee maker in the morning. Researching the coffee maker, you find that it uses 800 watts of power and you use it for one hour each day. So far your energy need is 800 watt-hours per day. You would then continue this process until you have accounted for all of your appliances and lighting.

To get a more precise measurement on the energy consumption of an appliance, you can also use a device such as the 'Kill A Watt', which will measure its actual usage.

The next factor is how much energy you are able collect, which is determined by the type of energy and your location. For instance, if you live in Arizona you will be able to collect the same amount of solar power per day with fewer collectors that someone living in Ontario.

Since this calculation is so heavily dependent on your location, it is best to use online resources which will take into consideration these various factors to help you determine your energy system's size. You can also see the section on resources in the back of this book for additional information.

Insulation provides a thermal barrier
between the living quarters of the
house and the outdoors

INSULATION
—

21

INSULATION TYPES

There are several different types of insulation options for a tiny house. The most common are closed cell spray foam (CCSF), extruded polystyrene (XPS) boards, and fiberglass insulation. Other less common options also exist including cotton denim, sheep's wool, as well as others.

The effectiveness of insulation is measured by an R-value, which is a unit of thermal resistance. The higher the R-value the better.

Material	R-value / Inch
Fiberglass Batting	3.14
Cotton Denim Batting	3.4
Sheep's Wool Batting	3.5
Open Cell Spray Foam	3.6
Extruded Polystyrene (XPS)	5
Closed Cell Spray Foam	6.5

CLOSED CELL SPRAY FOAM (CCSF)

CCSF is a great option for a tiny house because it has one of the highest R-values of any insulating material, at approximately 6.5R per inch. In addition to having a high R-value, CCSF is also very resistant to air leakage through the wall and roof cavity because it is sprayed in and expands to fill the area in which it is applied. This cuts down on drafts and improves the energy efficiency of the home. It also acts as a vapor barrier when applied to a minimum thickness of 2 inches, eliminating the expense and labor of adding one later. Finally, CCSF also becomes extremely ridged after it expands and dries, which can contribute to the structural strength of a tiny house.

The biggest disadvantage of CCSF is that it is expensive. CCSF is typically 3 times the cost of fiberglass insulation installed, or even more if the homeowner were to install the alternative themselves. The second disadvantage is that spray foam is not very do-it-yourself (DIY) friendly. While there are DIY kits, they can be more expensive than professional installation. These kits may also require expensive protective equipment that would only be used once. Additionally, unlike the temperature regulated rigs of the professionals, DIY kits can only be used within somewhat strict temperature ranges. If it is applied at the wrong temperature or incorrectly (e.g. two thick on a single pass), spray foam can outgas long after it has hardened. Unless there are no CCSF installation contractors in your area, I recommend hiring a professional to do this job.

EXTRUDED POLYSTYRENE (XPS) BOARDS SUPPLEMENTED WITH SPRAY FOAM

XPS boards are a good alternative to CCSF. They provide a relatively high R-value of approximately 5R per inch. They are considerably less expensive than CCSF, and if an

Closed Cell Spray Foam

XPS Board Insulation

Fender Area Partially Filled With Foam

aerosol foam product is used to fill the gaps between any adjoining boards and around the board edges, they still provide a good air barrier. They are also readily available at many home improvement retailers.

A disadvantage of XPS boards is that they can be somewhat difficult to install. Each board must be cut to fit within the wall cavity; the distance between the studs or rafters. Cutting these boards is a little more difficult than cutting wood, since they are not as rigid, and friction from tools like a circular saw can melt the product. If multiple layers of boards are used, these cuts will need to be made multiple times for each cavity. To cut thicker XPS boards, I recommend using a circular saw or handsaw. For thinner boards, a utility knife can be used to score it allowing for it to be easily and cleanly broken. Expect that the blade of the knife will need to be replaced frequently.

Common thicknesses of XPS boards are 2 inches and ¾ inch. Since it may be desirable to have more than 2 inches of foam in the wall cavity, a combination of multiple boards can be used to obtain the desired thickness. We do not recommend using a combination of boards that will exactly match the depth of a wall cavity. For instance, 3½ inches for a wall constructed of 2x4s. This is because XPS boards are very difficult to compress and if there are any size irregularities in the lumber or any obstructions in the cavity itself, the boards may extend out of the cavity and make it difficult to install the interior siding later.

FIBERGLASS

Fiberglass is the most common type of insulation used in residential construction. This is primarily driven by its low cost and easy installation. However, there are significant and numerous disadvantages to using fiberglass insulation in a tiny house. These include a relatively low R-value of 3.14R, when optimally installed. If the fiberglass batting becomes pinched, wet, or it settles, then the R-value significantly diminishes. Fiberglass batting is faced with kraft paper that can act as a vapor retarder, but it is not as effective as other products like plastic sheeting. Depending on your climate you may still need to add an additional vapor barrier. Like all batting, fiberglass also does not seal the wall cavity like expanded foam, so drafts from gaps are much more likely. Finally, fiberglass insulation requires the use of protective equipment during installation.

Since there are various disadvantages, I recommend against using fiberglass insulation in a tiny house.

OTHER INSULATION MATERIALS

There may be other insulation materials available depending on your location. These include effective natural options like cotton denim, sheep's wool, straw, and hemp. These typically have R-values similar to fiberglass, but they are all natural. If you have sensitivity to chemicals or would just prefer to have a natural insulation, these may be an option for you. Since they are not as common, you will likely need to special order them depending on your location.

The interior siding can set the tone
and style of a tiny house

INTERIOR
SIDING & TRIM

—

22

In typical residential construction, the most common type of interior siding is drywall. For a tiny house though, drywall is not recommended since it is both heavy and relatively brittle, especially at the joints. A more suitable material is either solid sheet paneling or tongue and groove (T&G) slat paneling. Solid sheet paneling is the easier of the two to install; however, some varieties can have a dated look and the solid sheets can have unattractive seams. Slat paneling, on the other hand, is more difficult and time consuming to install, but is often more attractive.

Trim Around Kitchen Window

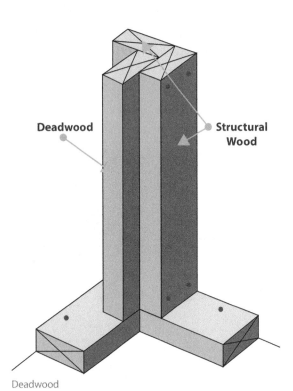

Deadwood

Structural Wood

Deadwood

Deadwood Installed in Ceiling Corner

Interior Pine Slat Siding

The downside to using paneling over drywall is that some adjustments may need to be made in various places to compensate for the differences in the thickness of the two materials. Most notably, these adjustments include changing electrical junction boxes so that they stand out beyond the studs only ¼ inch instead of ½ inch and adding ¼ inch of filler behind the window trim, since the window jamb also extends beyond the studs ½ inch.

Before any interior siding can be installed, there needs to be two pieces of framing wood in every corner that the siding can be nailed to. Often these boards will already be there as a result of the framing, but in some cases they will need to be added. These added boards are called either "nailers" or "deadwood" as they are only there to provide a surface to nail to and offer no structural support.

The siding should be unwrapped and given time to acclimate to the environment before being installed. The amount of time it takes can vary based on the specific product, but a minimum of 72 hours is recommended.

To attach the siding, use a pneumatic brad nail gun and construction adhesive. Adhesive can be applied to all the studs in a small area where the siding will be installed within approximately 10-15 minutes. Please note, this does not apply to fast drying/bonding adhesive. By pre-applying adhesive, you can avoid stopping to apply glue for each board. Be sure that the boards are tight against the studs wherever adhesive is used,

Trim Around Dormer Windows

since the boards will not be able to be made tighter after the adhesive dries.

The boards closest to the floor should be installed first, with the upper boards following. The first (bottom) board should be raised off the floor by a ½ inch to allow for expansion. This gap will not be visible later, as it will be hidden by the flooring and trim. The end of any board that will butt up to another board, not in a corner, should be cut at a 45 degree angle, in the same way as with the exterior siding. This allows the boards to overlap so that the studs behind the joints are not visible if the siding were to contract.

Once all the interior siding is installed, the trim can be added. The trim type to be used around the windows and doors is called casing. If the window trim goes all around a window, it is called "picture frame casing" because it looks like the window is framed. To install picture frame casing, you will measure the width and the height of the window you are framing. Be sure to take your measurement such that the trim edge will land approximately in the center of the jam. The width at the top of the window and the bottom of the window should be the same, as well as the heights on both sides. Once the measurements are taken, cut the casing with each end at 45 degrees. Next, on a large work surface, glue and nail the pieces together. Use a vice, if necessary, to ensure the corners are tightly assembled. Finally, install the trim as a single piece on the window. By assembling the trim as a single piece before installing, you ensure that the trim is square and that all the corners are tight.

For the remaining trim, cove or quarter round molding is used on the inside corners and outside corner molding is used on the outside corners.

When measuring and cutting trim, it is important to be very accurate because the trim is the final layer and there will be no more opportunities to cover up any imperfections. When measuring from one corner to another corner, it can be difficult to get an accurate measurement using a tape measure. This is because the tape measure must be bent and pushed into the second corner, making it difficult to read the exact measurement. To get around this, measure an arbitrary distance from the first corner and make a small pencil mark on the wall. Next measure from the second corner to the pencil mark and add the two numbers together. This can be made easier if your first measurement is a nice and easy round number like 10 or 20 inches.

Occasionally smaller pieces of trim will need to be cut. These pieces should be cut off of larger pieces, so that your hands can stay a safe distance from the cutting blade. If after cutting a small piece, you find that an additional cut is needed, switch to either a hand saw or cut a new piece from another longer piece. Do not attempt to hold a small piece of wood close to a power tool's blade with your hand.

Untrimmed Interior Siding Above Shower

Siding Around Switch

A durable, attractive material should
be chosen for the flooring

FLOORING

—

2

Hardwood Flooring

There are many different types of flooring that can be selected for a tiny house.

HARDWOOD FLOORING

Hardwood flooring is attractive, durable, and can flex with the trailer as required when the house is being towed. While hardwood flooring can be a more expensive option; with the limited amount of material required in a tiny house, this cost difference may not be that significant. Substantial deals on small quantities of flooring can often be found at flooring centers, since materials may be left over from larger jobs and product lines are occasionally discontinued with some stock remaining.

Engineered hardwood flooring has the same appearance as solid hardwood and is real wood, but is more stable and less susceptible to shrinking and expanding with changes in temperatures and humidity. Instead of being solid, it is layered much like plywood, with the top layer having the desired appearance. Engineered hardwood is also less expensive than solid hardwood. The disadvantage to engineered hardwood flooring is that it cannot be refinished as many times as solid wood.

Another option is laminate flooring. Laminate flooring is the least expensive 'wood look' option and can be very easy to install. However, laminate flooring generally does not look like real wood and may look cheap. Also, while laminate flooring is very durable, once it is damaged it cannot be repaired like wood flooring.

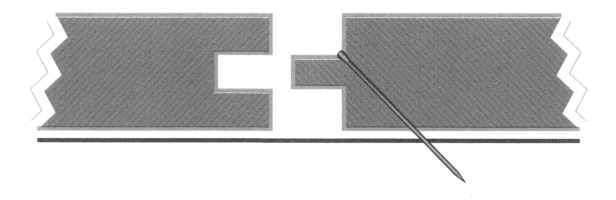

Blind Nail

Different flooring materials have different installation instructions so see the manufacturer for their specific directions. Below is a list of common steps to install hardwood flooring.

ACCLIMATE THE WOOD

As was done with the interior siding, allow the hardwood flooring to acclimate to the environment where it will be installed. The length of time that is recommended can vary from 72 hours to 14 days, so check with the manufacturer for details about your specific product. Be sure that the boxes containing the flooring are not sealed and avoid stacking them. Engineered wood generally has a much less recommended acclimation time.

INSTALLING THE FIRST ROW

The first row of hardwood flooring installed is the most important, as all subsequent boards will be based off of it. If this row is installed improperly, the entire floor will likely not look right.

The first decision when preparing to install the flooring is deciding on the direction of the boards. Boards can be installed so that they run parallel with the longer side of the room, parallel with the shorter side of the room, or at an angle. In a tiny house, it is best to install the boards parallel to the longer side of the house. This results in them being perpendicular to the joists, which adds strength to the floor, and is more visually appealing since it makes the house look larger.

Optionally, an underlayment can be laid out on the subfloor sheathing before installing any boards. This is not required, but can help reduce squeaks and act as a moisture barrier.

The first line of boards should start at the edge of the floor and be spaced approximately ½ inch away from the wall. This gap is to allow for expansion of the wood and will be hidden later by the wall trim. Measure and mark a line using a chalk line for the first row of boards. If the framing around the fenders is in the way of the chalk line, mark the closest line that can be continuous and then measure for the shorter lines on either side of the fender from the continuous line.

Additionally, double check your measurements and the line placement by verifying that the distance from the line to the opposite wall is the same on each end of the room. If it is not the same, then one of the walls may be slightly out of square and you may need to adjust your line so that the difference is split between both sides.

Attach the first row of boards by using a pneumatic nail gun to drive a nail at a 45 degree angle through the base of the tongue on the board. The nail should be driven in far enough so that it will not get in the way of the tongue and groove joining while installing the next row of boards. This process is referred to as blind nailing, since the nail will not be visible when the next row is installed. Be sure that the boards do not shift during this process. Additional nails can be driven into the top of the board, as long as they are close enough to the edge so that they will be covered by trim later. If they will not be covered then you will need to fill in the nail holes later with a color matched wood filler, but this should be avoided if possible.

INSTALLING MOST OF THE REMAINING BOARDS

To install the next row of boards, layout all the boards for that row ensuring to stagger the seams between rows in a natural pattern. Put each board in place and use a piece of scrap flooring against it and lightly hammer it to get a tight bond between the two rows. Be sure that the existing rows of boards do not move during this process. As additional rows are added, the chance of the existing boards moving reduces. Either use a hardwood flooring stapler or continue to blind nail the boards to the subfloor. Continue this process until all the boards are installed. When you get near the last row you will no longer have the room required to blind nail the boards.

INSTALLING THE LAST ROW

Since there will not be room for a piece of scrap wood or a hammer, use a flat bar placed against the wall to get a tight bond on the last row of boards. Be sure that the flat bar is against a solid surface, like a stud, otherwise it may break through the wall siding. Top nail the last row to attach it. As with the first row, a ½ inch gap should be left between the wall and the last row of flooring boards.

VINYL FLOORING

Hardwood flooring is not a good option in the bathroom because of the higher moisture levels found there. For this space, you will want to use a material that is more water resistant. In a typical home, this might be ceramic or stone tile, but because of their heavier weight they are not the best option for a tiny house. Vinyl tiles, however, have a similar appearance to ceramic tile, especially when they are grouted. They are also lightweight, strong, and inexpensive, making them a much better option for a tiny house.

INSTALLING VINYL TILES

When installing vinyl tiles, be sure that the arrows printed on the underside of the tiles all face in the same direction. These markings are added by the manufacturer to ensure the tiles fit together correctly.

Hard vinyl tiles can be cut by using a utility knife to score a line and then bending them until they snap. If a portion of a tile needs to be cut out, run the utility knife over the same area several times until it cuts all the way through the tile.

These tiles usually come with their underside already coated in an adhesive, but an additional layer of vinyl adhesive on the floor will ensure they stay in place. A trowel is used to apply the adhesive at the recommended thickness. The adhesive will need to fully dry before any tiles are placed on it. Once dry the adhesive becomes very tacky. If tiles are placed on the adhesive before it has a chance to dry, it will be squeezed up between the tiles, making for a difficult mess to clean up.

If the tiles are to be grouted, which is optional, use 1/16 inch spacers between the tiles to leave room for the grout. Use an unsanded grout because the gap between the tiles will be quite small.

TRANSITIONS

Add an additional piece of wood to cover the transition between hardwood flooring and other flooring types such as vinyl tiles. Different types of flooring have different thicknesses, so the transition piece has to be a different thickness on each side in order to rest flat when it is installed. Your flooring supplier should also carry transition pieces, with a matching finish to the hardwood floor. This piece is attached by top nailing it.

Installing the cabinets and shelving is one of the last steps in building a tiny house

CABINETS & SHELVING

—

Your choice for cabinets and their placement can heavily influence whether you are happy or disappointed with your house. They will determine how much storage you have, and to some degree, how roomy your house feels. It is important to have a place for each of your belongings in order for your house to stay organized and not feel cramped or messy.

For the cabinets in your house, you can either purchase pre-built cabinets, custom order them, or build your own.

Pre-built Cabinets Installed

PRE-BUILT CABINETS

Pre-built cabinets can be purchased from most home improvement stores. They are relatively cheap and come in an assortment of sizes. They are also easy to install and can be used to complete an entire kitchen in just a few hours.

The problem with pre-built cabinets is that, while they do come in an assortment of sizes, they may not have the right size for a tiny house. This is not as much of a concern for the lower cabinets, since you would not want to deviate from the standard countertop height which is comfortable to stand at, nor the standard depth which will properly fit a sink. However, for the upper cabinets, you may not have the room to place them as high as intended, especially if your kitchen is situated under a loft. Pre-built cabinets can be trimmed down to the correct depth in some cases, but the height cannot be altered. Any changes will weaken the cabinet box and will require reinforcement. To get around the issue of not being able to find upper cabinets that are the right size, many people instead choose to install open shelving.

The other issue that may come up with pre-built cabinets is the selection of the materials and finishes of the fronts and doors. This is usually fairly limited, since the stores need to stock so many sizes of each style variation. If you have an IKEA in your area, you may find a wider selection there. They are able to offer a wider selection because the fronts or 'faces' of their cabinets are sold independent of the backs or 'boxes', requiring some assembly.

To install pre-built cabinets, you will first need to join together any adjacent cabinets. This is done by lining up the framing on the face of the cabinets and clamping them together. You then pre-drill holes in the framing and screw the cabinets together. Once the cabinets are joined, they are screwed to the studs in the walls through the thicker pieces of wood on the back of the cabinets. Care should be taken to ensure that the cabinets are level when they are attached.

CUSTOM ORDER CABINETS

There are two different classifications of custom cabinets. There are those that are truly custom, which are constructed to the exact measurement for where they will be installed. Then there are those that are selected from a catalog of available sizes and made for you. This second type is what is found at the home improvement stores. Either way, custom order cabinets offer a much greater selection of sizes and styles. The downside is the cost. Custom order cabinets are considerably more expensive than any of the other options.

The home improvement stores generally offer free estimates for custom cabinets that include a 3D drawing of what the cabinets would look like once they are installed. Even if you do not plan to purchase the cabinets, laying out the design with a professional who has designed many kitchens can be a beneficial exercise. Your kitchen may be the smallest one they have ever seen, but their advice will still be valuable.

BUILD YOUR OWN

Building your own cabinets is the most economical choice and will result in cabinets that are perfectly sized for your needs. The challenge is that cabinets can be difficult to construct. The techniques to construct cabinets and furniture are very different than those used in house construction. While the box portion of a cabinet is easy to build, fronts require a different set of tools aimed primarily at joining wood together so there are no visible fasteners.

Woodworkers have different techniques to hide fasteners, but one of the most popular and easiest is pocket-hole joinery. Pocket-hole joinery is accomplished by drilling a hole into the back side of a piece of wood, usually at 15 degrees. A screw is then used to join it to another piece a wood. Using this approach, the fasteners are only visible from the back and will be hidden from view. In the case of furniture where the holes might be visible, wood inserts are available to fill and hide the holes.

What makes this approach to wood joining so easy is tools like the Kreg Jig®. Using this tool, you simply insert a piece of wood and it helps you to drill the pocket-holes at the correct location, angle, and depth. Then you line up the pieces of wood you are joining and screw them together. Select the correct length of fasteners to make sure you do not drive a screw out through the other side of the wood. The jig instruction will provide you with the correct fastener length for the size of wood you are working with.

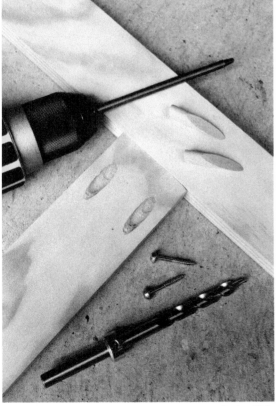

Pocket-Hole

Kreg Jig

CONCLUSION

There are many steps involved in building a tiny house. While many of these steps may take a long time to complete, they are generally not that difficult. Almost anyone willing to research, learn and try can build their own tiny house.

For each chapter in this guide, there are many books available that go into further detail on the subject. If you get to a part that is unclear, simply take your time and do some additional research. The answers are out there. If there is something you are just not comfortable doing yourself, like the plumbing or electrical installation, you can always hire someone to complete a specific task for you.

Starting a project like building an entire home from scratch can be very intimidating. While it may seem overwhelming, keep in mind that it is nothing more than a bunch of individual tasks that can be learned and mastered one at a time.

I hope this book has given you a little more knowledge, as well as a lot more confidence to start and complete your very own tiny house.

RESOURCES

For additional information and resources visit
https://www.tinyhomebuilders.com/resources

By putting the resources online we can ensure that they are always up to date with the latest information and trends.

CPSIA information can be obtained
at www.ICGtesting.com
Printed in the USA
LVOW05s0143280416

485458LV00004BA/4/P

9 780997 288704